Noise Pollution

Noise Pollution

Noel Templeton

R CALLISTO REFERENCE

www.callistoreference.com

Callisto Reference,
118-35 Queens Blvd., Suite 400,
Forest Hills, NY 11375, USA

Visit us on the World Wide Web at:
www.callistoreference.com

ISBN: 978-1-64116-624-9 (Hardback)

Cataloging-in-Publication Data

Noise pollution / Noel Templeton.
p. cm.
Includes bibliographical references and index.
ISBN 978-1-64116-624-9
1. Noise pollution. 2. Pollution. I. Templeton, Noel.
TD892 .N65 2022
620.23--dc23

Table of Contents

Preface

The propagation of noise which has harmful impacts on the animal life or human activity is known as noise pollution. Some of the common sources of noise pollution are machines, airplanes, loud music and construction. There are a lot of adverse health effects associated with it pertaining to both animals and humans. In animals, it can interfere with reproduction and navigation, and can also cause permanent hearing loss. In humans, noise pollution can increase the occurrence of coronary artery disease. Additionally, it can have serious adverse effects on the physical and psychological health of children. The set of strategies, which are aimed at lowering noise pollution or reducing the impact of noise falls under the area of noise control. The topics included herein on noise pollution are of utmost significance and bound to provide incredible insights to readers. This book is a compilation of chapters that discuss the most vital concepts in the field of noise pollution. It will provide comprehensive knowledge to the readers.

Given below is the chapter wise description of the book:

Chapter 1- The audible wave of pressure that propagates through a transmission medium such as solid, liquid and gas is referred to as sound. The harmful impact on activities of human and animal life caused by the propagation of noise is called noise pollution. This is an introductory chapter which will briefly introduce all the significant aspects of noise pollution as well as some metrics for noise assessment.

Chapter 2- Transportation noise is the noise emitted by various means of transport. It has many auditory and non-auditory effects on human health. It can also be categorized into roadway noise, railway noise and aircraft noise. The topics elaborated in this chapter will help in gaining a better perspective about transportation noise.

Chapter 3- Industrial machinery and processes are composed of various noise sources such as rotors, stators, gears, fans, vibrating panels, turbulent fluid flow, electrical machines, internal combustion engines, etc. This chapter closely examines the different sources of industrial and construction noise to provide an extensive understanding of the subject.

Chapter 4- There are various impacts of noise pollution on human health and the environment. Some of these impacts include cardiovascular disease, hypertension, sleep disturbance, cognitive impairment in children, etc. This chapter has been carefully written to provide an easy understanding of these impacts of noise pollution.

Chapter 5- Many mitigation approaches are used for noise control and management. A few of these approaches and techniques are urban planning, noise control engineering, architectural acoustics, soundproofing, etc. This chapter closely examines these key aspects associated with noise management and mitigation to provide an extensive understanding of the subject.

Chapter 6- Acoustic instruments are used in measuring and assessing the sound levels by the use of sound pressure. Sound level meter or decibel meter is one such instrument used for sound level measurement. The topics elaborated in this chapter will help in gaining a better perspective about the concepts of sound intensity and noise calculation as well as instruments associated with it.

At the end, I would like to thank all those who dedicated their time and efforts for the successful completion of this book. I also wish to convey my gratitude towards my friends and family who supported me at every step.

Noel Templeton

Chapter 1

Noise Pollution: An Introduction

The audible wave of pressure that propagates through a transmission medium such as solid, liquid and gas is referred to as sound. The harmful impact on activities of human and animal life caused by the propagation of noise is called noise pollution. This is an introductory chapter which will briefly introduce all the significant aspects of noise pollution as well as some metrics for noise assessment.

Sound

Sound is a pressure wave which is created by a vibrating object.

These vibrations set particles in the surrounding medium (typical air) in vibrational motion, thus transporting energy through the medium. Since the particles are moving in parallel direction to the wave movement, the sound wave is referred to as a longitudinal wave.

The result of longitudinal waves is the creation of compressions and rarefactions within the air.

The alternating configuration of C and R of particles is described by the graph of a sine wave (C ~crests, R ~troughs).

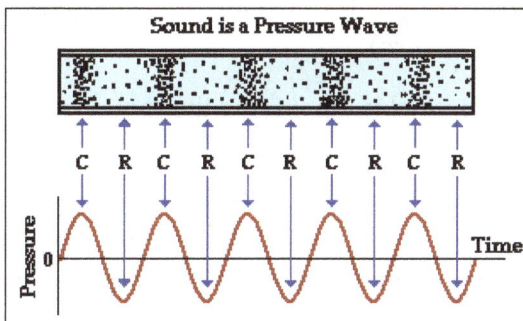

Sound is a Pressure Wave

The speed of a sound pressure wave in air is $331.5 + 0.6T_c$ m/s, T_c temperature in Celsius. The particles do not move down the way with the wave but osciallate back and forth about their individual equilibrium position.

Wavelength, Amplitude and Frequency of a Wave

The amount of work done to generate the energy that sets the particles in motion is reflected in the degree of displacement which is measured as the amplitude of a sound.

The frequency f of a wave is measured as the number of complete back-and-forth vibrations of a particle of the medium per unit of time. 1 Hertz = 1 vibration/second f = 1/Time.

Depending on the medium, sound travels at some speed c which defines the wavelength l: l = c/f.

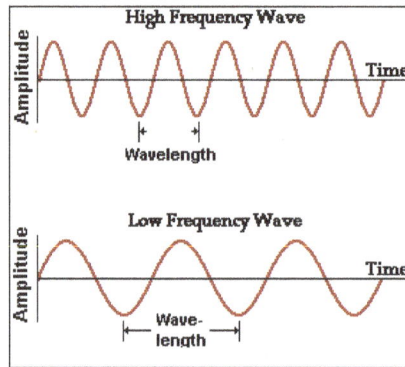

Measuring the Intensity of Sound

- The softest audible sound modulates the air pressure by around 10^{-6} Pascal (Pa). The loudest (pain inflicting) audible sound does it by 10^2 Pa.

- Because of this wide range it is convenient to measure sound amplitude on a logarithmic scale in Decibel [dB].

- Decibel is not a physical unit - it expresses only a ratio for comparing the intensity of two sounds: $10 \log_{10} (I/I_0)$ where I and I_0 are two intensity/power levels ($I \sim P^2$, P is sound pressure).

- One can say e.g. a channel is amplifying the sound by 3 dB, meaning the output is 3 dB louder than the input.

- In order to make it interpretable as a real unit, a fixed pressure $P_0 = 2*10^{-5}$ Pa is defined (the reference of 0db corresponds to the threshold of hearing) and the absolute sound pressure P in Decibel is defined as: $20 \log_{10} (P/P_0)$.

- Thus +20 dB means an increase in pressure by a factor of 10.

Table: Examples for Sound Levels in Decibel.

Threshold of hearing	0 dB	softest audible 1000 Hz sound	6 dB
quiet living room	20 dB	soft whispering	25 dB
refrigerator	40 dB	soft talking	50 dB

normal conversation	60 dB	busy city street noise	70 dB
passing motorcycle	90 dB	somebody shouting	100 dB
pneumatic drill	100 dB	helicopter	110 dB
loud rock concert	110 dB	air raid siren	130 dB
pain threshold	120 dB	gunshot	140 dB
rocket launch	180 dB	Instant perforation of eardrum	160 dB

1) TOH: One-billionth of a centimeter of molecular motion.

2) The most intense sound (without physical damage) is one trillion times more intense.

Noise Pollution

Noise pollution is an unwanted or excessive sound that can have deleterious effects on human health and environmental quality. Noise pollution is commonly generated inside many industrial facilities and some other workplaces, but it also comes from highway, railway, and airplane traffic and from outdoor construction activities.

Measuring and Perceiving Loudness

Sound waves are vibrations of air molecules carried from a noise source to the ear. Sound is typically described in terms of the loudness (amplitude) and the pitch (frequency) of the wave. Loudness (also called sound pressure level, or SPL) is measured in logarithmic units called decibels (dB). The normal human ear can detect sounds that range between 0 dB (hearing threshold) and about 140 dB, with sounds between 120dB and 140 dB causing pain (pain threshold). The ambient SPL in a library is about 35 dB, while that inside a moving bus or subway train is roughly 85 dB; building construction activities can generate SPLs as high as 105 dB at the source. SPLs decrease with distance from the source.

The rate, at which sound energy is transmitted, called sound intensity, is proportional to the square of the SPL. Because of the logarithmic nature of the decibel scale, an increase of 10 dB represents a 10-fold increase in sound intensity, an increase of 20 dB represents a 100-fold increase in intensity, a 30-dB increase represents a 1,000-fold increase in intensity, and so on. When sound intensity is doubled, on the other hand, the SPL increases by only 3 dB. For example, if a construction drill causes a noise level of about 90 dB, then two identical drills operating side by side will cause a noise level of 93 dB. On the other hand, when two sounds that differ by more than 15 dB in SPL are combined, the weaker sound is masked (or drowned out) by the louder sound. For example, if an 80-dB drill is operating next to a 95-dB dozer at a construction site, the combined SPL of those two sources will be measured as 95 dB; the less intense sound from the compressor will not be noticeable.

Frequency of a sound wave is expressed in cycles per second (cps), but hertz (Hz) is more commonly used (1 cps = 1 Hz). The human eardrum is a very sensitive organ with a large dynamic range, being able to detect sounds at frequencies as low as 20 Hz (a very low pitch) up to about 20,000 Hz (a very high pitch). The pitch of a human voice in normal conversation occurs at frequencies between 250 Hz and 2,000 Hz.

Precise measurement and scientific description of sound levels differ from most subjective human perceptions and opinions about sound. Subjective human responses to noise depend on both pitch and loudness. People with normal hearing generally perceive high-frequency sounds to be louder than low-frequency sounds of the same amplitude. For this reason, electronic sound-level meters used to measure noise levels take into account the variations of perceived loudness with pitch. Frequency filters in the meters serve to match meter readings with the sensitivity of the human ear and the relative loudness of various sounds. The so-called A-weighted filter, for example, is commonly used for measuring ambient community noise. SPL measurements made with this filter are expressed as A-weighted decibels, or dBA. Most people perceive and describe a 6- to 10-dBA increase in an SPL reading to be a doubling of "loudness". Another system, the C-weighted (dBC) scale, is sometimes used for impact noise levels, such as gunfire, and tends to be more accurate than dBA for the perceived loudness of sounds with low frequency components.

Noise levels generally vary with time, so noise measurement data are reported as time-averaged values to express overall noise levels. There are several ways to do this. For example, the results of a set of repeated sound-level measurements may be reported as L_{90} = 75 dBA, meaning that the levels were equal to or higher than 75 dBA for 90 percent of the time. Another unit, called equivalent sound levels (L_{eq}), can be used to express an average SPL over any period of interest, such as an eight-hour workday. (L_{eq} is a logarithmic average rather than an arithmetic average, so loud events prevail in the overall result). A unit called day-night sound level (DNL or L_{dn}) accounts for the fact that people are more sensitive to noise during the night, so a 10-dBA penalty is added to SPL values that are measured between 10 PM and 7 AM. DNL measurements are very useful for describing overall community exposure to aircraft noise, for example.

Dealing with the Effects of Noise

Noise is more than a mere nuisance. At certain levels and durations of exposure, it can cause physical damage to the eardrum and the sensitive hair cells of the inner ear and result in temporary or permanent hearing loss. Hearing loss does not usually occur at SPLs below 80 dBA (eight-hour exposure levels are best kept below 85 dBA), but most people repeatedly exposed to more than 105 dBA will have permanent hearing loss to some extent. In addition to causing hearing loss, excessive noise exposure can also raise blood pressure and pulse rates, cause irritability, anxiety, and mental fatigue, and interfere with sleep, recreation, and personal communication. Noise pollution control is therefore of importance in the workplace and in the community. Noise-control ordinances and laws enacted at the local, regional, and national levels can be effective in mitigating the adverse effects of noise pollution.

Environmental and industrial noise is regulated in the United States under the Occupational Safety and Health Act of 1970 and the Noise Control Act of 1972. Under these acts, the Occupational Safety and Health Administration set up industrial noise criteria in order to provide limits on the intensity of sound exposure and on the time duration for which that intensity may be allowed.

If an individual is exposed to various levels of noise for different time intervals during the day, the total exposure or dose (D) of noise is obtained from the relation

$$D = (C_1/T_1) + (C_2/T_2) + (C_3/T_3) + ... + (C_n/T_n),$$

where C is the actual time of exposure and T is the allowable time of exposure at any level. Using this formula, the maximum allowable daily noise dose will be 1, and any daily exposure over 1 is unacceptable.

Criteria for indoor noise are summarized in three sets of specifications that have been derived by collecting subjective judgments from a large sampling of people in a variety of specific situations. These have developed into the noise criteria (NC) and preferred noise criteria (PNC) curves, which provide limits on the level of noise introduced into the environment. The NC curves, developed in 1957, aim to provide a comfortable working or living environment by specifying the maximum allowable level of noise in octave bands over the entire audio spectrum. The complete set of 11 curves specifies noise criteria for a broad range of situations. The PNC curves, developed in 1971, add limits on low-frequency rumble and high-frequency hiss; hence, they are preferred over the older NC standard. Summarized in the curves, these criteria provide design goals for noise levels for a variety of different purposes. Part of the specification of a work or living environment is the appropriate PNC curve; in the event that the sound level exceeds PNC limits, sound-absorptive materials can be introduced into the environment as necessary to meet the appropriate standards.

Low levels of noise may be overcome using additional absorbing material, such as heavy drapery or sound-absorbent tiles in enclosed rooms. Where low levels of identifiable noise may be distracting or where privacy of conversations in adjacent offices and reception areas may be important, the undesirable sounds may be masked. A small white-noise source such as static or rushing air, placed in the room, can mask the sounds of conversation from adjacent rooms without being offensive or dangerous to the ears of people working nearby. This type of device is often used in offices of doctors and other professionals. Another technique for reducing personal noise levels is through the use of hearing protectors, which are held over the ears in the same manner as an earmuff. By using commercially available earmuff-type hearing protectors, a decrease in sound level can be attained ranging typically from about 10 dB at 100 Hz to more than 30 dB for frequencies above 1,000 Hz.

Outdoor noise limits are also important for human comfort. Standard house construction will provide some shielding from external sounds if the house meets minimum standards of construction and if the outside noise level falls within acceptable limits. These limits are generally specified for particular periods of the day—for example, during daylight hours, during evening hours, and at night during sleeping hours. Because of refraction in the atmosphere owing to the night time temperature inversion, relatively loud sounds can be introduced into an area from a rather distant highway, airport, or railroad. One interesting technique for control of highway noise is the erection of noise barriers alongside the highway, separating the highway from adjacent residential areas. The effectiveness of such barriers is limited by the diffraction of sound, which is greater at the lower frequencies that often predominate in road noise, especially from large vehicles. In order to be effective, they must be as close as possible to either the source or the observer of the noise (preferably to the source), thus maximizing the diffraction that would be necessary for the sound to reach the observer. Another requirement for this type of barrier is that it must also limit the amount of transmitted sound in order to bring about significant noise reduction.

Metrics for Assessing Environmental Noise

Selecting a metric for assessment of environmental noise is no simple task, because it must reflect the impact on people. No single metric can describe all responses in all situations. Context, expectations, and people's experiences and circumstances all affect their responses. Hence, levels of community response (e.g., annoyance) may vary from community to community, just as individual responses vary from person to person, even if noise levels do not change. However, one consistent finding has been that changes in noise exposure do affect individual and community responses and that increases in man-made noise usually have a negative impact.

Thus, it is important to understand which characteristics of noise elicit a negative response and how exposure to noise with those characteristics affects people's lives. The metric chosen or developed for measuring community noise must reflect this human response and must be taken into account in making policy decisions.

Fifty years ago, when noise metrics were developed, the choices were based on simpler calculations and technologies and the acoustical quantities that could be predicted by sound propagation models used at the time. Although much more sophisticated measurements can be made today, many still consider these "older" metrics valid and continue to use them. However, with modern instruments, much more accurate measurements and predictions can now be made of people's reactions to noise.

A meaningful metric, or set of metrics, translates sound pressure-time history measurements into a prediction of the effects of noise, such as annoyance, sleep disturbance, changes in health, interference with understanding speech, and ability to learn. Ideally, this translation should be based on context, expectations, and personal situations and preferences, in addition to noise information, and should account for a distribution of responses, including responses of vulnerable populations, such as children. Unfortunately, a holistic model of community response is still beyond present capabilities.

One fundamental issue that must be considered in the choice of an environmental noise metric(s) is the purpose for which the metric will be used:

- To implement public policy on noise emission from one or more sources.

- To provide information about noise exposures in a form understandable to the public.

- To assess a noise situation in terms of noise control engineering.

The metrics to accomplish these purposes may differ, but all three relate directly to the impact of noise on the community. For example, a metric to inform decisions about noise control engineering strategies should result in reducing the noise impact, which would then be reflected in the policy metric(s) and the public information metric(s).

As new research results become available and accessible, they should influence the choice of metrics for the three purposes listed above. The results of such research may result in complex calculations that include many variables and may better quantify individual reactions to sound. Some modern procedures, such as calculation of loudness, are more complex than earlier methods, but available computational procedures make the results widely available.

Loudness and A-weighting

Arguably the modern history of noise metrics began in the 1930s with the search for a way to describe the loudness of sound. This led to the definition of weighting networks for sound-level meters and, because of limitations on the capabilities of calculating sound pressure levels at that time, a single frequency-weighted value—either A-weighted or C-weighted—came into common usage.

Loudness

In an early attempt to determine the loudness of sound (using discrete-frequency tones), Fletcher and Munson found that the loudness of a tone depends on both its amplitude and its frequency. Knowing this dependence, they were able to develop a set of equal-loudness curves. In modern terms the unit of loudness is the phon. For example, a 1,000-Hz tone with a sound pressure level of 40 dB has a loudness of 40 phon. At this loudness level the sound pressure level of tones between 1,000 and about 5,000 Hz is generally lower than 40 dB, and the sound pressure level of tones below 1,000 Hz and above about 5,000 Hz is higher than 40 dB.

The sound-level meter was standardized in the early 1930s when microphones and electronic circuits were being developed. Ideally, the standard sound-level meter would have a single-number description of the sound at a given point in space. The best description at the time came from the studies by Fletcher and Munson, who clearly showed that the shape of the equal-loudness curve was dependent on both the amplitude and the frequency of sound. Thus, using the linear electronic circuits of the time, a few curves had to be selected based on the amplitude of the sound. One of the curves selected, which is very close to the 40-phon curve, was designated as "A-weighting". Another, which was nearly independent of frequency, was designated as "C-weighting". A third curve, the "B-weighting" curve, which fell between the A and C curves, has long since fallen out of favor. A-weighting and C-weighting are still used today, although the shape of the curves has changed somewhat to provide a standardized mathematical description in terms of poles and zeros of a transmission network.

Work on improving the calculation of loudness based on measurement of the spectrum of sound continued. The best-known early work in the United States was by S. S. Stevens and in Germany by Eberhard Zwicker. Stevens's Mark VI and Zwicker's work on loudness were standardized by the International Organization for Standardization. Later work by Brian Glasberg and Brian Moore in the United Kingdom was the basis for the American National Standard on computation of loudness.

Over the years, A-weighted levels were found to correspond reasonably well to human response, especially for noise spectra in typical offices. Single-number methods of rating noise in offices and other building spaces were also developed, including so-called noise rating curves (NR curves—a curve tangent method of obtaining a single number from an octave band spectrum) and ratings based on loudness and A-weighting.

Metrics for Measuring Community Reaction to Noise

One early attempt to develop a metric for forecasting community response to noise was made by Stevens. Unlike the DNL, this metric included nonacoustical factors as well as noise levels and yielded a "composite noise rating". This rating was then plotted against a scale of community responses—vigorous community action, threats of community action, widespread complaints, sporadic complaints, and no observed reaction. A few case studies showed a reasonable correlation between the measurement and response but with considerable scatter. Community noise levels were determined by measuring the average octave band levels in the community averaged in space and time. A curve tangent method was used to reduce the octave band data to a single-number rating.

Day-night Average Sound Level

After EPA established the Office of Noise Abatement and Control and after passage of the Noise Control Act of 1972, EPA was faced with the task of developing a metric for community noise with the following characteristics:

- The measure should be applicable to the evaluation of pervasive long-term noise in various defined areas and under various conditions over long periods of time.

- The measure should correlate well with known effects of the noise environment on the individual and the public.

- The measure should be simple, practical, and accurate. In principle, it should be useful for planning as well as for enforcement or monitoring purposes.

- The required measurement equipment, with standardized characteristics, should be commercially available.

- The measure should be closely related to existing methods currently in use.

- The single measure of noise at a given location should be predictable, within an acceptable tolerance, from knowledge of the physical events producing the noise.

- The measure should lend itself to small, simple monitors that can be left unattended in public areas for long periods of time.

EPA also published its rationale for choosing A-weighting and for leaving open the possibility of using a different metric in the future:

> With respect to both simplicity and adequacy for characterizing human response, a frequency-weighted sound level should be used for the evaluation of environmental noise. Several frequency weightings have been proposed for general use in the assessment of response to noise, differing primarily in the way sounds at frequencies between 1000 and 4000 Hz are evaluated. The A-weighting, standardized in current sound level meter specifications, has been widely used for transportation and community noise description. For many noises, the A-weighted sound level has been found to correlate as well with human response as more complex measures, such as the calculated perceived noise level or the loudness level derived from spectral analysis. However, psychoacoustic research indicates that, at least for some noise signals, a different frequency weighting which increases the sensitivity to the 1000–4000 Hz region is more reliable. Various forms of this alternative weighting function have been proposed; they will be referred to here as the type "D-weightings". None of these alternative weightings [have] progressed in acceptance to the point where a standard has been approved for commercially available instrumentation.
>
> It is concluded that a frequency-weighted sound pressure level is the most reasonable choice for describing the magnitude of environmental noise. In order to use available standardized instrumentation for direct measurement, the A-frequency weighting is the only suitable choice at this time. The indication

that a type D-weighting might ultimately be more suitable than the A-weighting for evaluating the integrated effects of noise on people suggests that at such time as a type D-weighting becomes standardized and available in commercial instrumentation, its value as the weighting for environmental noise should be considered to determine if a change from the A-weighting is warranted.

The decision to add 10 dB in measuring night time levels and the selection of a two-period (day-night) metric rather than a three-period metric (day-evening-night) was based on community reaction studies at the time and tests that showed little difference between a two-period and a three-period metric. Thus, the DNL (A-frequency weighting for both day-time and night time levels and a 10-dB increase in measuring system gain at night) came into being for the evaluation of community noise.

In the United States, DNL and the percentage of persons highly annoyed are widely used, especially by the Federal Aviation Administration (FAA). The Federal Highway Administration uses A-weighting and the average sound pressure level during the busiest traffic hour as a measure of community impact. The difference between C-weighted and A-weighted levels is used as an indication of the low-frequency content of the sound, and the sound exposure level is used to evaluate sounds of finite duration—for example, an aircraft flyover.

Day-evening-night sound level is widely used in Europe. In some countries, L_{day} and L_{night}, (average A-weighted sound pressure levels) are used in addition to or instead of a DNL-type metric. None of these metrics takes into account the time of night when the noise occurs, even though noise appears to cause greater sleep disturbance at the beginning and end of the night.

Several issues have arisen from the use of DNL and the percentage of persons highly annoyed: no one actually "hears" a DNL; there is a high variability from study to study around a nominal Schultz curve; and in many situations "highly annoyed" is not an appropriate measure of human response. Although the percent highly annoyed and DNL approach has been widely endorsed, variability around a nominal Schultz curve is troubling, and there are reports that this approach is not sufficient to predict community response. Attitudinal and personal variables impact people's responses and are, to some extent, the reason for scatter.

Some researchers have found in their analyses of survey results that the nominal Schultz curve appears to depend on the noise source (e.g., aircraft, road traffic, rail traffic). In addition, DNL is a relatively insensitive measure of sleep disturbance and thus is not an appropriate metric for predicting awakenings in sleep disturbance studies. Finally, A-weighting is not the best weighting for measuring noises with unusual spectra (e.g., excessive high- or low-frequency noise or noise that has unusual peaks in its spectrum). For sounds with levels that evolve over time, the most appropriate weighting should change with the level; typically, however, only one weighting is used.

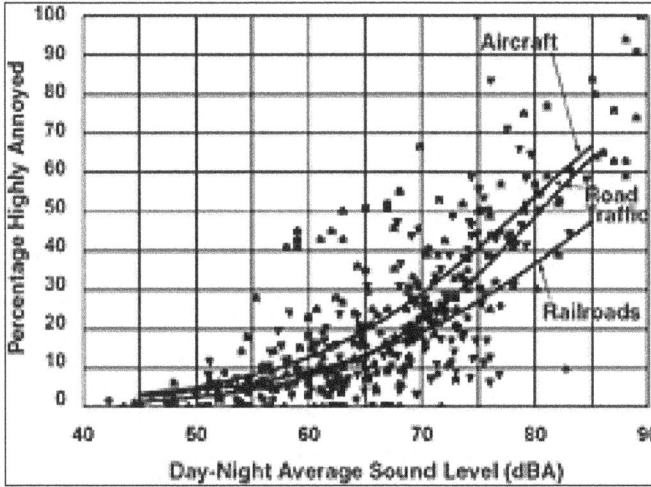

Variability in survey results. ▼ = road traffic. ✷ = air traffic. ◆ = rail traffic.
Curves are the results of fits to data associated with different modes of transportation.

Percentage of Persons Highly Annoyed

The next major event in the selection of a noise metric was a study by Schultz of surveys of community reaction to noise. Schultz went back to original data to estimate the percentage of the population "highly annoyed" as a function of DNL. Even at that time, it was recognized that, for a variety of reasons, there was considerable scatter in the data. Nevertheless, Schultz proposed that a single curve (the Schultz curve) drawn through the data should be used as a measure of community response. Later studies led to modifications of the Schultz curve. In the latter study, three curves were compared, and a U.S. Air Force logistic curve was defined

$$\%HA = 100/[1+ \exp (11.13 - 0.14L_{dn})]$$

The scatter in the highly annoyed response, compared to scatter in the average curve, was presented by Miedema and Vos and has been commented on by several subsequent researchers. The first problem with scatter is that it causes great uncertainty in the prediction of community reaction. A second problem is that community reaction (percent highly annoyed) appears to depend on the source of the noise; for example, responses to aircraft noise, road traffic noise, and rail noise vary, even if the noises have the same DNL. The question that must be answered is whether the variability in response is due to the nature of the noise source or reflects how the metric is calculated.

Consultants and other professionals are often asked to study community noise issues and recommend remedial action. Predictions of community response should not be based only on variations of the Schultz curve. It has been known for many years that nonacoustical factors influence community reaction to noise. Thus, at a minimum, temporal and spectral variations must also be taken into account.

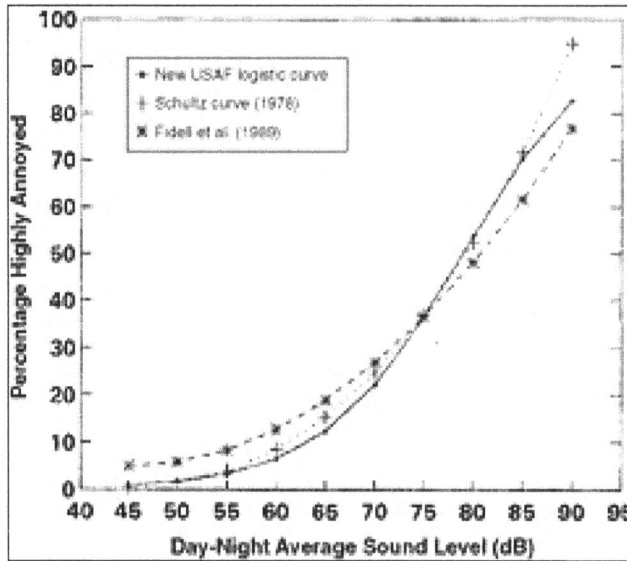

Three versions of a Schultz curve. ■ = the U.S. Air
Force logistic curve. = the curve proposed by Schultz. * = a curve.

Based on work by EPA, Schomer proposed modifications to DNL to account for tonality, impulsiveness, background noise, type of community, and other factors. Schomer also showed how this modified approach could be used to reduce variances in the survey data on which the Schultz curve is based.

The Federal Interagency Committee on Noise endorsed the use of percent highly annoyed and DNL as metrics for assessing community noise around airports and recommended that the equation above be accepted as showing the definitive relationship between percent highly annoyed and DNL. Response curves for community annoyance have now been standardized nationally and internationally.

Alternative Metrics

The science of measuring environmental noise has progressed rapidly in the past decade as computer technology has come on line to provide rapid data acquisition and analysis in small portable packages. The end result has been a revolution in the type and complexity of measurements and calculations that can be made in analyzing environmental noise. This topic provides a more detailed description of presently used metrics and a variety of alternative metrics that are well within the capabilities of modern instrumentation.

A Different Frequency Weighting

An alternative to A-weighting (i.e., D-weighting) could be considered. As noted earlier, this weighting was considered by the EPA in 1974 but rejected because there was no standard shape for the curve.

Perceived Noise Level

Community reaction to noise from jet planes led to important events in the development of noise metrics. The problem, which arose in 1956, is described in an autobiography by Beranek. According to measurements made with a standard sound-level meter, the noise produced by a Boeing 707 jet airplane and that by a propeller airplane (Super Constellation) were equal. However, subjective testing showed that the 707 was considered much noisier; by subjective measures, the A-weighted sound pressure levels of the 707 would have to be significantly reduced to be considered as noisy as the Super Constellation. This early test of the usefulness of A-weighted levels in judging noisiness led to further evaluations of the relative noisiness of propeller-driven and jet airplanes and the development of the concept of "perceived noisiness".

Perceived noise level (PNL) was used in the development of specifications of noise emissions from airplanes for regulatory purposes in 1969 and is still used to certify airplanes today. When the perceived noise-level metric was adopted, it was possible to compute it only with a large amount of equipment. Today, it can be done with a hand-held sound-level meter. D-weighting simplifies the PNL calculation, but neither PNL nor D-weighting solves the decibel issue, which relates to explaining noise to the public.

Loudness

Historically, the method of calculating PNL was similar to the method of calculating loudness. Today, several methods can be used to calculate loudness, all of them with a hand-held sound-level meter. Loudness that exceeds some agreed-on value a given percentage of the time also can be calculated. On a linear scale (as opposed to a logarithmic scale), a doubling of the value of the calculated value corresponds to a doubling of the loudness. This may be easier to explain to the public than a metric that uses the phon (which uses a logarithmic scale) as a unit of loudness. For sounds in the midfrequency range, an increase in A-weighted level of 10 dB corresponds to a doubling of loudness.

Speech Interference

Standard methods of calculating speech interference are available, and the values may be translated into effects that are easier for the public to understand than DNL. For example, the difficulty of communicating over a given distance between speaker and listener may be quantified in terms of percentage of speech likely to be understood. Speech interference can be affected by the fact that hearing loss increases with age and usually starts at high frequencies. Thus, the ability to distinguish consonants that have high-frequency content such as "s" and "th" is diminished.

Nighttime Sleep Disturbance

In Night Noise Guidelines for Europe, published by the World Health Organization, sleep disturbance is related to the night time level designated as L_{night}, although researchers also use indoor LAmax and indoor A-weighted sound exposure level (ASEL) when investigating the relationship between awakenings and noise. The temporal pattern of noise at night, however, is known to influence sleep disturbance. This problem is addressed to some extent in a new American National Standard in which terms such as the likelihood of awaking, are used; the new standard may be more understandable to the public than the day-night average level or the nighttime level used in Europe.

Metrics for Communicating with the Public

An often-cited shortcoming of DNL is that the public does not understand what it means. Over the years, various people have advocated using supplemental metrics that describe noise in ways that are more understandable to the majority of people. Metrics used to supplement DNL include time above (a certain level), number of events above a given value of the ASEL, number of loud events above a certain ASEL in a given period, and single-event descriptors such as L_{Amax} and ASEL. Most advocate using a group of metrics to give a fuller picture of the potential impact of the exposure and explaining that these measures supplement metrics such as DNL. The same argument can be made for using a group of metrics when addressing other measurements or predicting a variety of impacts, such as the number of occurrences of speech interference; when noise levels inside buildings exceed recommended levels for a particular activity, such as learning in schools; or the likelihood of being awakened based on predicted indoor single-event metrics.

The number of events has been recognized as an important factor in noise exposure, and it is included in metrics that are or have been used to predict annoyance; alternatives to DNL, such as the Noise Exposure Forecast (NEF) system used in Canada and elsewhere; and the Noise and Number Index (NNI) that was used prior to 1990 in the United Kingdom. The NEF metric is based on effective perceived noise level as well as the number of events; hence it takes into account some of the impact of tonal components and impulsiveness on annoyance. NNI is also based on a very basic loudness measure, perceived noise level in decibels and number of events. Analysis of data from a study at U.K. airports in 1982 and another study in 2005 showed that the relationship between annoyance and A-weighted equivalent level had changed. However, by combining a measure of average noise exposure with the number of events, it was possible to develop a metric that worked consistently for both studies.

Noise Metrics for Rural/Naturally Quiet Areas

Neither day-night average sound level nor percent highly annoyed is an appropriate metric for measuring noise in naturally quiet areas. Because of the logarithmic nature of the decibel, short-duration sounds of high amplitude compared with background noise can significantly increase the day-night level, even though the sound remains at the background level most of the time. As for percent highly annoyed, this is hardly the best measure of satisfaction for areas where quiet and solitude are valued. In addition, it can be difficult to measure very low sound pressure levels. A-weighted levels of 40 dB are at the upper end of the range, and lower levels can be at or even below the levels measurable with conventional sound-level meters.

Nevertheless, some quantification of noise impact is clearly needed in these areas as a basis for establishing public policy, which usually means regulatory action. The classic definition of noise is "unwanted sound," so the source of sound must be identified, either as part of the natural soundscape or not. Thus, simple metrics like sound pressure level are clearly not appropriate. For example, an airplane overflight may have a much lower sound pressure level and shorter duration than sound from a rushing stream, but the former is considered noise and the latter is considered sound. The method of assessment of the noise environment should also take into account the likely long-term impact on animals that use, for example, very low level sounds (perhaps inaudible or unnoticed by people) to locate prey or predators.

International Activities Related to Noise Metrics

The International Commission for the Biological Effects of Noise holds meetings at five-year intervals. In 2008 the meeting was held in the United States, but most of the participants came from other countries, as did the presenters. Truls Gjestland of Norway presented a summary report on research in the past five years related to the effects of community noise, specifically annoyance. Although some research has been done in Japan, he said, not many significant projects had been undertaken. However, he noted that at least three different versions of the Schultz curve had been developed, all of them based generally on the same datasets. Around the same time Lawrence Finegold of the United States presented a review of major noise-related policy efforts around the world during the same time period.

European Activities

In 1996 the European Union (EU) published The Green Paper, which established new noise programs that are used to address noise issues today. European Directives have been issued concerning noise emissions from consumer products, and an EU Environmental Noise Directive (END) in 2002 led to the development of noise-mapping and, in a few cases, action plans that require noise metrics. Related activities include the HARMONOISE and IMAGINE projects.

European Metrics (Indicators)

A-frequency weighting for determining sound levels that have been standardized in the United States and internationally is widely used in Europe. However, as discussed elsewhere in this chapter, frequency weighting alone is not enough to define a metric. A Working Group (WG1) that produced a report in 2000, Position Paper on EU Noise

Indicators, in support of future European noise policies, identified five criteria for selecting an indicator: validity, practical applicability, transparency, enforceability, and consistency. Although this report was not an official EU document, the metrics recommended therein are now widely used.

WG1 recommended that two indicators, both based on A-frequency weighting, be used for reporting data on noise exposure. These indicators were designated L_{EU} and $L_{EU}N$ but today they are widely known as the day-evening-night sound level, DENL, and the equivalent sound pressure level during the eight-hour nighttime period, L_{night}. The group explained, and questioned, the rationale for using 5 dB as level weighting for the evening period and 10 dB for nighttime. Nighttime was nominally designated as eight hours, from 11:00 p.m. to 7:00 a.m.; daytime, 12 hours; and evening, four hours (with some variation, depending on the country). For general purposes, the long-term average A-weighted sound pressure level, L_{Aeq}, was used. The WG1 report also recognized that the character of noise (impulsive, tonal, etc.) may affect human response. Thus, corrections to the metrics may be necessary, and A-frequency weighting may not be appropriate for measuring low-frequency noise.

The WG1 report was also the basis for metrics specified in the 2002 END that led to noise mapping. The directive also suggests supplemental metrics based on the WG1 report.

Dose-effect Relationships (Exposure-Response Relationships)

Another Working Group (WG2) on Health and Socio-Economic Aspects of Noise also produced a report, again not official EU policy. In Position Paper on Dose-Response Relationships between Transportation Noise and Annoyance, the group recommended that the percent highly annoyed (%HA) be used as a measure of community response to noise. Updated and modified Schultz curves, based on the work of Miedema and colleagues for the %HA as a function of day-evening-night sound level, are used to measure road traffic, rail traffic, and aircraft noise. WG2 also acknowledged the variability from study to study in the mean values in Schultz curves but still supported the use of "norm" curves.

Substantial deviations from the predicted percentage of annoyed persons must be expected for limited groups at individual sites because random factors, individual and local circumstances and study characteristics affect the noise annoyance. However, in many cases the prediction on the basis of a "norm" curve that is valid for the entire population is a more suitable basis for policy than the actual annoyance of a particular individual or group. For example, a "norm" curve is useful when exposure limits for dwellings and noise abatement measures are discussed. Equity and consistency require that limits and abatement measures do not depend on the particularities of the persons and their actual circumstances. For similar reasons, a "norm" curve also can be used to estimate the number of annoyed persons in the vicinity of an airport, road, or railway when different scenarios concerning, e.g., extension of these activities or emission reductions are to be compared. That the norm curve does not take local circumstances or reactions to a change in exposure itself into account, is considered to be an advantage for many purposes. Equity and consistency of policy would not be served if in each case the actual annoyance is taken as the (only) basis for these evaluations. The use of "norm curves" or "norm thresholds," which are valid for the entire population (or a particular sensitive subgroup), is common practice when exposures to other environmental pollutants, such as air pollutants or radiation, are evaluated. There they are used for the evaluation of an individual situation, irrespective of the population in that situation. It is recommended to take the same approach in the case of environmental noise and use the same curve irrespective of the population in the situation evaluated.

Nighttime Effects

In 2004, WG2 produced Position Paper on Dose-Effect Relationships for Nighttime Noise, again not an official EU document. In this paper the metric used was Lnight, as defined above as the measure for sleep disturbance. Based on questionnaires, curves similar to Schultz curves were developed, the ordinate being the percent highly disturbed and the abscissa being the nighttime noise level. An effort was made to relate single events to the nighttime sound level.

Annoyance and the Microstructure of Noise Exposure

Several studies have been published, mostly in connection with the EU-funded SILENCE project, on the importance of the microstructure of a noise exposure situation. The argument is that equivalent levels do not "tell the full story". Different traffic noise situations with the same equivalent level may be assessed differently with respect to annoyance. This is important information for decisions about how to reduce the negative impact of road noise through traffic management measures. Laboratory experiments have provided several examples:

- An even flow of traffic causes the same annoyance as when vehicles are clustered, but an even flow is more damaging to mental performance than clustered traffic.

- A large difference between equivalent level and L_{max} is more annoying than a small difference.

- Trams should receive a 3-dB "bonus" over buses.

- Different noises from a rail yard at equal equivalent levels may have a subjective difference of as much as 5 dB.

Recommendations for Future Research in Europe

Research for a Quieter Europe in 2020, a report produced under the auspices of the CALM Network, provides a strategy for future noise research in the EU. The report includes an excellent review of EU activities related to noise and covers a wide variety of future needs, including noise emissions from various sources and the need for perception-based research into the effects of noise.

European versus Japanese Results on Transportation Noise

A recent Japanese study by Yano compared the effects of transportation noise in Japan with the EU results. The effects of road traffic noise are similar, but the effects of railway noise were quite different. The authors suggest that the differences may be attributable to the proximity of Japanese homes to railroad tracks (where they are subject to vibration as well as noise). Differences in the construction of homes may also be a factor.

Japanese data for aircraft noise are based on one dataset of 410 responses around Kumamoto, a small airport, and may not be representative of noise around Japanese airports in general. There was also an active anti-noise group near Kumamoto airport. However, considering the scatter from study to study, the results of the Kumamoto study may be representative.

Summary Findings and Recommendations

Established and New Environmental Noise Metrics

Use of the DNL metric has helped policy makers, road planners, airport managers, the public, and others understand potential noise impacts on communities and have helped guide noise mitigation efforts around airports, roadways, and rail systems. However, DNL has both strengths and weaknesses as a measure of noise.

The strengths of DNL are that it has become familiar over time, its calculation has been standardized, through experience it has become well understood, and it is now embedded in software used for planning. DNL has made it possible to communicate evaluations of noise to the public to provide people with a better understanding of how noise policy decisions are made and how changes in transportation systems, or choosing to live near an airport or a busy highway, might affect them. DNL has also been a mechanism by which people could be protected and systematically helped to address problems with environmental noise exposure fairly and equitably.

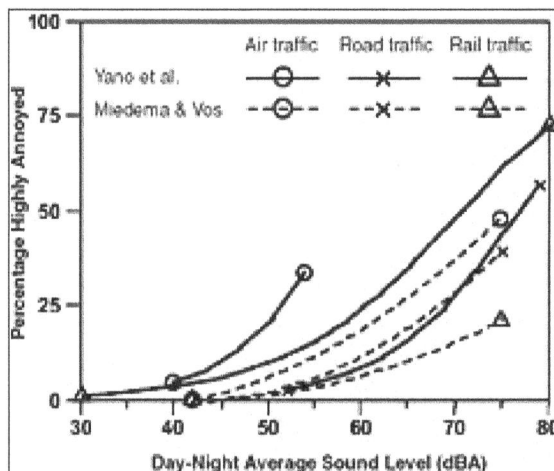

Comparison of the present dose-response curves with results from Miedema and Vos.

DNL also has some drawbacks. First, there is a great deal of variability from study to study in the percentage of the population believed to be "highly annoyed" as a function of DNL, which predicts only part of a community's response to noise. Efforts to develop metrics that can provide a more definitive assessment of community impact are still a topic for research and policy debate.

Many limitations of a DNL-type metric based on the average A-weighted sound pressure level used to assess environmental noise have been noted:

- DNL is insensitive to the impact of very loud, isolated events.

- Fewer loud events can have the same DNL as many quieter events; thus, the impacts of very different soundscapes are described as equal.

- DNL is insensitive to the time when an event occurs (e.g., noise early in the night causes different sleep disturbance than noise early in the morning).

- The only strong argument for using night and evening weightings in DNL is based on the fact that average nighttime ambient levels are lower than those during the day.

- Other metrics such as speech interference level and nighttime levels provide a better measure of annoyance with speech interference and conscious awakenings.

- DNL is an outdoor noise measure that may not reflect differences between outdoor sounds and the same sounds heard indoors.

- A-weighting does not reflect the results of research studies in psychoacoustics over the past 40 years.

- DNL does not take into account other sound characteristics (e.g., tonality and rate of loudness onset) that can influence annoyance and sleep disturbance levels.

Although DNL has limitations, it has served as the major environmental noise metric since the early 1970s. Despite the variability in community response, it is clear that the percentage of the population highly annoyed for a DNL of 65 dB is considerably greater than the corresponding percentage for a DNL of 55 dB. This supports the findings of EPA in the 1970s that a DNL of 55 dB is the level necessary to protect the public health and welfare with an adequate margin of safety.

When new metrics are developed and values selected as a matter of public policy, the goal should be to protect a larger fraction of the population than is protected under the value now widely used—the DNL = 65 dB criterion. Many steps would have to be taken before a different metric (or set of metrics) could be recommended to policymakers. Changing to another metric would entail significant effort and cost (e.g., in conducting surveys and development of databases) and would be of limited value unless the new metric offers significant benefits over DNL, most importantly in providing a more transparent and definitive connection between noise level and annoyance or other effects on people's lives. Unfortunately, because of a lack of "real-world" data to test the performance of metrics, it is difficult to establish their advantages and disadvantages. The situation with respect to DNL has been recognized by the FAA, and two meetings have been held—one in August 2009 and one in December 2009—to discuss a "roadmap" to improve the situation regarding noise metrics.

A set of metrics, rather than a single metric, to describe different types of outcomes of environmental noise (e.g., number of interruptions of speech, learning impairment in schools, number of additional awakenings) would provide a multidimensional picture of noise impact and may be the best approach to informing the public. Supplementary metrics could make possible predictions of noise from transportation in sufficient detail to enable the development of noise maps.

When communicating with the public, it might be useful to translate metric values into words (e.g., categories such as no observed reaction, sporadic complaints, widespread complaints, threats of community action, vigorous community action) that can be more easily understood than DNL and other numerical metrics.

The ability to predict direct health effects of noise (e.g., hypertension, speech interference, cognitive impairment, sleep disturbance) and the relationship between these effects and annoyance requires further study in order to develop new metrics that account for health effects.

The federal government (e.g., agencies of the U.S. Department of Transportation with responsibilities related to noise and the U.S. Department of Housing and Urban Development) should adopt as a goal the 1974 recommendation of the Environmental Protection Agency (EPA, 1974) to limit the day-night average sound level (DNL) to 55 decibels (dB) to protect the public health and welfare. Currently, DNL (DENL in Europe), the accepted metric for characterizing the impact of community noise, shows that a large proportion of the population is highly annoyed at a DNL of 65 dB or higher.

Relevant agencies of the federal government (e.g., agencies of the U.S. Department of Transportation with responsibilities related to noise, the Environmental Protection Agency, and the U.S. Department of Housing and Urban Development) should fund the development of environmental noise metrics that are more transparent and more reflective of the impact of noise on an affected population than DNL. This will require improved tools for predicting community sound pressure time histories and the development of metrics that accurately reflect the sounds people hear. A more holistic model of annoyance is also needed that incorporates situational variables that can be used to generate predictions for overall response, as well as responses of vulnerable populations (e.g., elderly people, sick people, children, and noise-sensitive individuals). International cooperation in this effort will facilitate the development of national and international standards for calculating metrics and should include open-source code to facilitate broad implementation of the metrics. Certain measures should be taken to facilitate this development:

- The international noise control engineering community should develop an open, collaborative data-sharing environment in which researchers can deposit and access data from community noise surveys (e.g., data from surveys of acoustic, environmental, community, and transportation systems to support comparisons of metrics and predictions by models).

- Policy agencies should conduct extensive surveys around at least six U.S. airports to generate high-quality data to populate the database. These surveys should serve as models of good survey practices, including data recording and archiving to ensure that they are useful for future studies.

Noise Metrics for Quiet Environments

The impact of man-made noise in national parks and other quiet environments is another parameter that is not well modeled by the metrics used to assess the impact of noise around airports or roads. Detection of the sound and distinguishing between man-made and natural sounds are important because human reactions to man-made and natural sounds differ. If one goal of the national parks is to preserve places of natural beauty, then the natural soundscape of a park, which is an aspect of its beauty, should also be preserved.

In addition, predicting the impact of noise on wildlife in national parks may require a different kind of metric that reflects animals' hearing systems. Preserving wildlife is essential to preserving the ecostructure of a park. But wildlife preservation will require that animals' hearing also be preserved and protected, because many animals depend on their hearing to hunt and to detect potential predators. The U.S. Department of the Interior should fund the development of metrics to support noise management decisions in national parks and other quiet environments.

Chapter 2

Transportation Noise

Transportation noise is the noise emitted by various means of transport. It has many auditory and non-auditory effects on human health. It can also be categorized into roadway noise, railway noise and aircraft noise. The topics elaborated in this chapter will help in gaining a better perspective about transportation noise.

Mobility is a lynchpin of our economy and a basic human need. But it can also be harmful to the environment and human health in many different ways.

Steadily growing levels of traffic, particularly in the commercial transport sector, in all probability cancel out the savings that can potentially be realized from improved engine technologies and the use of alternative fuels. Traffic generates not only greenhouse gases, but also particulate matter and NOx, which constitute a serious health hazard. It also generates noise, which at high levels is disturbing or even unhealthy for many people, not to mention its negative impact on quality of life.

The fact that traffic noise occurs virtually everywhere and all the time is mainly attributable to increased transport, plus the growing number of noise pollution point sources such as lawn blowers and large outdoor events. Another factor is that many people have become much more susceptible to environmental pollution, particularly when it comes to noise. Noise pollution from cars, trains and aircraft can only be substantially reduced by implementing a broad and harmonized range of measures involving vehicle and road technologies, tax regulations, and traffic and urban planning.

A comprehensive sustainable mobility plan would need to (a) prioritize persuading all concerned to use eco-friendlier means of transportation; and (b) include emission reduction measures that use latest generation technologies. Such measures should aim to promote the use of quiet, low-emission vehicles, low-emission driving techniques, and eco-friendly driving routes. Supremely important in this regard is adjusting emission limits to today's advanced technologies.

Roadway Noise

In most countries, the control of road vehicle noise as it affects the external observer (i.e., a person outside the vehicle and not its occupant) is a subject of legislation.

Although there is not much legislative control of the interior noise in the vehicles, such control is important too, since high noise levels are both objectionable and tiring to the majority of drivers and passengers.

From this point of view, control of interior noise in a vehicle should be an essential feature of its design. Unfortunately, the extent to which this (i.e., reduction of interior noise) is carried out depends, to a large degree, on the "commercial" considerations, i.e., how far a low interior noise level is a selling point for that particular class and design of the vehicle.

This is but natural under the present circumstances, since there are hardly any legal constraints, in most countries, on the interior noise levels in vehicles.

It is not necessarily true, however, that a programme of noise reduction from the "interior" or occupancy point of view will automatically ensure a low level of "exterior" noise.

It is quite possible, by a suitable use of sound-absorbing materials, and isolation and suppression of vibration sources, to design and produce a vehicle which is extremely quiet for its occupants, but still has excessive airborne exhaust noise, engine noise, or even road noise to the exterior listener.

In general, however, one may expect that reduction of exhaust noise to low enough levels (so that it is masked by the engine noise) will meet all legal requirements. In addition, this will also classify the vehicle concerned as "quiet" or "reasonably quiet" from the point of view of the external listener.

It is usually possible to achieve such a degree of silencing without any appreciable reduction in power. The only exception where practical difficulties may arise in this approach is the case of some sports cars with small ground clearance. In this case, accommodation of the necessary silencer volume creates problems.

Sources of Vehicle Noise

The engine is the main source of noise in road vehicles. Engine noise consists of intake noise, exhaust noise and the noise produced in the engine itself. Other sources of road vehicle noise and vibrations, along with their cause and treatment. Table gives a summary of resonance parameters, most likely causes of excitation and their treatment.

Table: Noise and vibration sources in road vehicles, their cause and treatment.

Source	Cause	Treatment
Gearbox	(1) Gear tooth contact	Usual method to reduce machine noise
	(2) Casing resonance	Use of stiff casing

Planetary gears	(1) Gear tooth contact	Difficult to cure, if present
	(2) Gear displacement	Difficult to cure, if present
Fluid coupling	(1) Blade frequency noise	Design problem
	(2) Cavitation	Increasing the static operating pressure
Power steering	Hydraulic noise	Application of decoupling
Rear axle	Gear tooth contact	Design problem

Prop shaft	(1) Unbalance	Removal of design or production fault
	(2) Poor alignment	Design problem
	(3) Poor coupling characteristics	Design Problem
	(4) Mechanical damage	Replacement of damaged parts (e.g. bent shaft)
Universal joints	Asymmetric motions	Preference for constant velocity joints
Body	(1) Cavity resonance	Sound deadening (stiffening is less effective)
	(2) Impact	Sound deadening and isolation
	(3) Relative movement	Design or production problem
Heater	(1) Motor noise	Isolation or enclosure; use of quieter motor and fan
	(3) Dust noise	Usually masked by motor/fan noise
Air conditioning	(1) Compressor noise	Isolation of enclosure; removal from passenger area
	(2) Pressure fluctuations	Use of silencers as necessary
Door	(1) Door shutting	Better design, local sound insulation and sealing
	(2) Lack of rigidity	Better door and body rigidity
Aerodynamics	(1) Wind flutter	Use of draught deflectors
	(2) Overall turbulence	Improvement of aerodynamics shape
	(3) Local turbulence	Improvement of aerodynamics of detailed design
	(4) Edge tones	Improvement of aerodynamics of detailed design
Road noise	(1) Tyre characteristics	Many factors are involved
	(2) Road surface	Related to tyre characteristics and suspension
Tyres	(1) Causing vibration	Numerous factors involved
	(2) Tyre thump	Elimination of discontinuities in tyres
	(3) Tyre squeal	Related to rubber composition road surface, steering geometry type of load, inflation pressure, etc.
	(4) Unbalace	Correction of wheels balance and regular rebalancing

Some of the producers of noise in vehicles can be treated at the source by balancing, damping, silencing, etc. The effect of the remaining noise and vibrations can be reduced, on the other hand, by isolation and damping, as far as is practicable. It may be noted here that isolation is effective for transmitted noise, while damping reduces radiated noise.

Another important point to be kept in mind in this connection is that the solutions of the problem of vehicle noise may not always be straightforward, but depend very much on empirical work. This is due to the fact that a major source of noise may very likely be an unanticipated resonance.

Table: Resonance as a source of road vehicle noise.

Parameter	Most-likely cause of excitation	Treatment
Bonnet-cavity resonance	(1) Induction noise (often to firing frequency) (2) Engine noise	Adequate intake silencer and bonnet lining Better design of engines
Body-cavity resonance	(1) Road noise transmitted through suspension (2) Engine noise	Sound deadening by proper lining Isolation of engine and firewall insulation
Boom resonance	(1) Silencer-body resonance (2) Engine noise (3) Road noise (4) Transmission flexure	Suitable size and ranges of silencer; sound deadening coating; floor carpeting Isolation of engine; sound-deadening linings, isolation of body-work Sound-deadening linings and coating; isolation of suspension and body-work Uniform stiffness

Engine Noise

Engine is the main source of noise in road vehicles. The effects of engine noise may be reduced by isolation. This is done by setting the engine unit on resilient mounts. Unfortunately, complete isolation by this method is impossible.

The degree of resilience necessary for complete isolation of the engine unit would result in a mounting which would be much too flexible. This problem is complicated still further by the fact that the support to which the resilient mounts are attached is itself relatively flexible and may itself be excited by road noise, for example.

The support may also be excited by the feedback of its own excitation from the un-sprung mass of the suspension. Thus it becomes necessary for the designer to work out the best possible compromise.

Since no straightforward solutions may be available, such a compromise has to be based initially on previous experience and such empirical data as are available. The design is subsequently adjusted or modified as required in the light of experimental testing and final evaluation under true road conditions.

The purpose of damping engine vibrations to a suitable degree by resilient mounting is to inhibit the transmission of engine noise to other parts of the vehicle, and also to eliminate the effects of suspension "shake" which can be feedback through the engine mounts. It is essential to maintain satisfactory isolation over a wide range of engine speeds, road speeds and operating conditions.

Such a degree of isolation involves adjustment of stiffness in all three planes and consideration of all possible modes of vibration of the engine so that resonance with the frequencies of other disturbances is avoided. It has been found that, in practice, the engine mount resonance usually lies between 5 Hz and 15 Hz.

The former frequency corresponds to the maximum flexibility which can be permitted without excessive static deflection of the engine under torque reaction.

Even when isolation of the engine has been achieved to a good extent, engine and induction noise may still be objectionably high due to bonnet-cavity resonance. This is treated by a sprayed-on sound-deadening lining to the underside of the bonnet. This treatment has become now a commonly adopted standard practice.

It may also be extended down the sides of the bonnet cavity, although such regions are generally stiffer and less likely to resonate at engine and induction noise frequencies.

Now a days, conversion "kits" are also available for reducing the noise produced by bonnet-cavity resonance. These kits comprise of sound-deadening "blankets" or linings, and are based on the same principle of reducing or eliminating bonnet-cavity resonance and acting as a barrier for radiated sound.

Isolation and Insulation of Vehicle Noise

The isolation and insulation of the engine. Isolation may also be extended to various points on the suspension and transmission. This table gives the relevant details of isolation treatment for various components (other than the engine) of a road vehicle. It is difficult to estimate the effectiveness of such treatment except experience.

Moreover, the damping effects may not necessarily be cumulative. The solution of this problem depends essentially on the proper design of the vehicle concerned as well as the quality of the road surface over which the vehicle is to run. Thus the problem of isolation is an individual one.

Table: Isolation treatment of a road vehicle.

Component	Parameters (natural frequency)	Remarks or treatment
Engine mount	(1) General noise reduction (below 15 Hz)	Critical for adequate performance
	(2) High-frequency noise (below 15 Hz)	-
	(3) Torque reaction (below 5 Hz)	Needs evaluation in detail
	(4) Engine Vibration noise (10-12 Hz)	-
	(5) Suspension shake (10-12 Hz)	-
Wheels	Vibrations transmitted from tyres to body	(1) Rubber bushing at front suspension pivot points. (2) Isolation mounts (3) Reduction of unsprung weight
Body	Vibrations transmitted to body	(1) Rubber mounts between body and frame (2) Sandwich panel construction

Insulation of damping, on the other hand, is more generally applicable and more consistent in response, as in the case of bonnet-cavity resonance. In the same manner, body-cavity resonance can be treated with linings for the sides, roof and floor.

Table: Sound Insulation treatment of a road vehicle.

Area	Treatment of	Typical materials
Bonnet lining	Bonnet-cavity resonance	Sprayed on coatings, blankets, etc.
Firewall	Engine noise	Felt mat, glass fibre mat, etc.
Floor	Engine and road noise, body cavity resonance	Carpet and felt underlay
Roof	Body-cavity resonance	Lining with or without additional absorbing layers
Sides	Body-cavity resonance	Trim panels with or without sprayed on coating
Doors	Body-cavity resonance	Trim panels with or without sprayed on coating, draught-excluding rubber seals (act as isolating dampers)
Underseal	Impact noise, engine noise, road noise, body resonance	Sprayed on thick plastic coating (limited sound deadening effect)
Seats	General noise level	Padded upholstery

Besides eliminating resonance, insulation treatment is especially effective in reducing road noise. In this respect, thick under sealing treatment rates high. In general, the larger the sheet metal area concerned, the greater the noise reduction likely to be achieved by insulation treatment.

This general rule applies particularly to lower-cost cars, since there is a tendency to use thinner gauge metal work for body construction of such cars. Moreover, insulation is more likely to be effective for the reduction of noise produced by resonance than local stiffening, as the latter may merely shift the point of vibration or resonance to another area in a unitary construction.

Boom Frequencies

Even when a more or less complete treatment using isolation and insulation methods has been applied to a road vehicle, certain "boom" frequencies may still remain. These frequencies are usually restricted to a particular engine speed or road speed.

In the case of engine speed, the boom frequencies are excited by the engine noise or, more usually, the exhaust noise, such boom frequencies may arise only because a part of the exhaust system is resonating under those particular conditions.

In some of such cases, the cure may simply be a matter of improving the isolation of the exhaust hangers. In other more complicated cases, however, more elaborate treatment may be necessary.

On the other hand, boom frequencies generated at particular road speeds usually mean that the body cavity resonance requires further damping treatment. Such frequencies may be initiated by suspension shake, resonant flexural vibration of the transmission, tyre noise or other similar causes. Boom frequencies related to particular road speeds require both isolation and damping treatments.

Noise Spectrum in Vehicles

Inside the vehicle, the sound spectrum is a characteristic "print" of the design and construction of the vehicle. It is related also to the isolation and insulation treatments involved. As a general rule, the overall noise level increases with speed almost linearly in the absence of marked resonance effects.

At higher speeds, however, a sharp increase in noise level can be anticipated unless the aerodynamic shape of the vehicle is satisfactory.

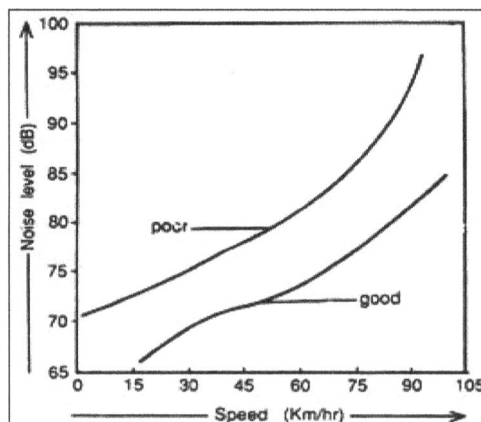

Typical variation of the overall noise level of vehicles with poor and good insulation as a function of their speed.

It should be noted that the difference between the noise levels corresponding to poor and good insulation vehicles can be as high as 5-10 dB between the corresponding curves.

In actual practice, deviation from the almost linear curve may be expected. There may also occur one or more apparent "peaks" of reasonably low order. Adequate insulation treatment will only ensure that the overall noise levels are not objectionable.

After this, there may not be much further improvement. Individual sharp peaks, on the other hand, may require a detailed investigation in order to affect a possible cure. For example, these sharp peaks may be due to suspension shake, tyre resonance, structural resonance, body cavity resonance or boom resonance.

The performance of a road vehicle is considered satisfactory in a majority of cases if no such sharp peaks are found in the noise spectrum when the vehicle is running on a normal, smooth road surface. On the other hand, if the vehicle is running over an inferior road surface, a marked increase in noise with sharp peaks at particular speeds may be expected.

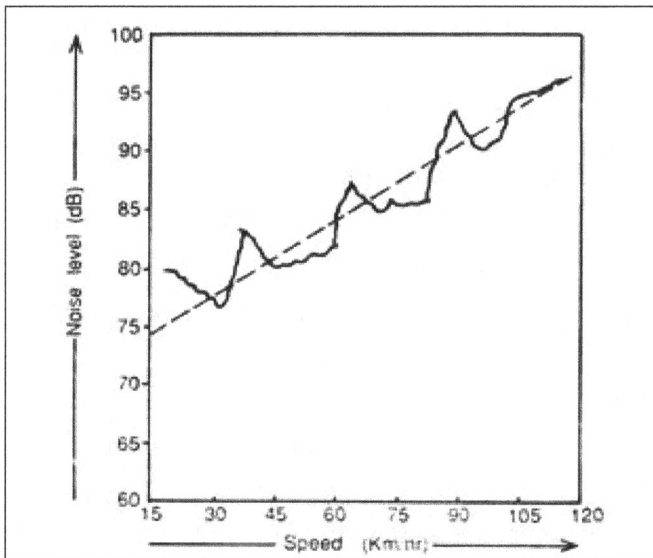

The overall noise levels of a vehicle as a function of its speed.

Of course, the noise spectrum also varies with the speed of the vehicle and the road surface conditions. The noise spectrum therefore, is specific to a particular vehicle running at a particular speed over a particular road surface. In general, the bulk of the noise content for a road vehicle running at moderate speeds on a reasonably smooth road surface is likely to be concentrated around 50 Hz or even lower frequencies.

With modern suspensions and current sound-insulation treatment, the actual noise levels at these frequencies will rarely be objectionable. These noise levels, in fact, may pass unnoticed. In sharp contrast to this, the road noise generated by more irregular road surfaces may show up as peaks at 200-600 Hz. In this case, the change in noise will immediately become noticeable and, in general, objectionable.

Brake Squeal Noise in Vehicle

Brake squeal is an annoying noise on some vehicles. The cause of excitation of this type of noise usually remains indeterminate and its effects are variable. The brake squeal may develop after some period of use. It may persist on one vehicle but remain completely absent from an identical model.

As regards the frequency spectrum of squeal, the frequencies may range from 2,000 Hz upwards and well into the ultrasonic range; but the most annoying frequencies usually lie between 10 and 15 kHz. If the vehicle has drum brakes, it is commonly believed that the most likely cause of break squeal is the accumulation of dust.

It has been observed, however, that complete cleaning of the accumulated dust will not necessarily clear the squeal. On the other hand, one brake with a given accumulation of dust may not squeal, whilst another does.

The most positive "cut-and-dried" method to cure brake squeal is to change the lining material. This cure, however, is far from infallible. In the case of disc brakes, a cure can often be achieved by using a suitable lubricant which does not result in the loss of braking efficiency. In many cases, brake squeal is self-curing in the sense that it disappears after a while and then it may or may not return.

Road Traffic Noise Calculation

The quality of human life, in fact, is heavily influenced by a continuous exposure to acoustical noise exceeding a threshold, usually defined in the dedicated country regulation or in the International Standards.

Therefore, the evaluation of Environmental Noise Impact due to acoustical noise has to be performed. This can be achieved both by a measurement campaign or by a software simulation. The latter needs a very precise mathematical modelling of the environment, of the sources and of the propagation law of sound.

The development of models to predict traffic noise started more than 50 years ago and, the results have often been very accurate. Usually these kind of models are developed taking into account mainly traffic flow, both of light and heavy vehicles, features of the road surface, distance between carriage and receivers. Moreover, since several models have been developed all around the world, the peculiarities of different countries, in terms of roads, kind of vehicles and weather features have to be taken into account.

The aim of using a Traffic Noise Model (TNM) is twofold: on one side it can be used in the designing of new road infrastructures in order to evaluate the acoustical impact and to avoid postconstruction mitigation actions that often present a greater cost; on the other side it can be used on an existing road network, so that the measurement campaign can be minimized and can be used just for the tuning of the model.

Many countries decided to regulate the use of these models, establishing which one can be adopted in a traffic noise simulation.

Most used Traffic Noise Models

Basic Statistical Models

First attempts of making a traffic noise prediction can be collocated into 1950/1960 decades; they mainly evaluate the percentile L_{50}, defined as the sound level exceeded by the signal in 50% of the measurement period.

These models refer principally to a fluid continuous flux, considering a common constant velocity with no distinction between vehicle typologies.

One of the first models, developed in 1952, is the one reported in Handbook of Acoustic Noise Control. This model states that the 50 percentile of traffic noise for speed of 35-45 mph (about 55-75 Km/h) and distances greater than 20 feet (about 6 meters) is given by:

$$L_{50} = 68 + 8.5\,Log\,(Q) - 20\,Log\,(d)$$

where Q is traffic volume in vehicles per hour and d is the distance from observation point to center of the traffic lane, in feet; no specification is included about vehicles and roads type.

In the following years, Nickson presented a new model in which a new parameter is included to relate the model with the experimental data. They proposed:

$$L_{50} = C + 10\,Log\left(\frac{Q}{d}\right)$$

where C is a constant value that can be evaluated making an analysis of experimental data and L_{50} is the sound level in dBA.

Later, Johnson presented a new TNM taking also into account the mean speed of vehicles in mph, v. They proposed for L_{50} the following expression:

$$L_{50} = 3.5 + 10\,Log\left(\frac{Qv^3}{d}\right).$$

This model presents a good agreement with the experimental data for a percentage of heavy vehicles from 0% to 40%. It also includes some corrective factor for ground attenuation and gradient.

Some years later, Galloway et. al improved this model taking into an account the percentage of heavy vehicles P. Their expression for the L_{50} level in dBA was:

$$L_{50} = 20 + 10 \, Log\left(\frac{Qv^2}{d}\right) + 0.4 \, P$$

The models developed in the next years introduced the equivalent level L_{eq} as sound level indicator. One of the most used is the Burgess Model applied for the first time in Sydney in Australia. Using the same notation of the previous expression, the sound level is given by:

$$L_{eq} = 55.5 + 10.2 \, Log(Q) + 0.3 \, P - 19.3 \, Log(d)$$

Another most used calculation formula is called "Griffiths and Langdon Method". In particular they propose the evaluation of equivalent level starting from the percentile level as follow:

$$L_{eq} = L_{50} + 0.018\left(L_{10} - L_{90}\right)^2$$

where the statistical percentile indicator have the expression:

$$L_{10} = 61 + 8.4 \, Log(Q) + 0.15 \, P - 11.5 \, Log(d)$$
$$L_{50} = 44.8 + 10.8 \, Log(Q) + 0.12 \, P - 9.6 \, Log(d)$$
$$L_{90} = 39.1 + 10.5 \, Log(Q) + 0.06 \, P - 9.3 \, Log(d)$$

where Q, P and d have the same meaning of previous formula.

Several years later Fagotti improved the previous models introducing the motorcycles and buses flux, Q_M and Q_{BUS}. The formula they propose is the following:

$$L_{eq} = 10 \, Log\left(Q_L + Q_M + 8Q_P + 88Q_{BUS}\right) + 33.5$$

Another model was formulated by the French "Centre Scientifique et Technique du Batiment" (C.S.T.B.), which proposed a predictive formula of equivalent emission level, based on the average acoustic level (L_{50}) with the following expression:

$$L_{eq} = 0.65 \, L_{50} + 28.8 \, [dBA]$$

The value of L_{50} is calculated taking into account only the equivalent vehicular flows (Q_{eq}), and is given by:

$$L_{50} = 11.9 \, LogQ + 31.4 \, [dBA]$$

For urban road and highway with vehicular flows lower than 1000 vehicles/hour;

$$L_{50} = 15.5 \, LogQ - 10 \, LogL + 36 \quad [dBA]$$

For urban road with elevated buildings near the carriageway edge, with L the width (in meters) of the road near the measurement point.

It is easy to notice that all the previous models can be deducted by the general expression of the equivalent level calculated according to a statistical traffic noise model is:

$$L_{eq} = A \cdot LogQ \left[1 + \frac{P}{100}(n-1) \right] + b \cdot Log(d) + C$$

Since a heavy vehicle generates a stronger noise than a light one, a factor n, called acoustical equivalent of heavy vehicles, has been considered. Therefore an equivalent traffic flow, Q_{eq}, can be formulated as follows:

$$Q_{eq} = Q \left[1 + \frac{P}{100}(n-1) \right]$$

The A, b and C coefficients may be derived, for a fixed investigated area, by linear regression methods on many L_{eq} data taken at different traffic flows (Q, P) and distances (d). The acoustical equivalent, n, (defined as the number of light vehicle that generate the same acoustic energy of an heavy one) can be estimated both by regression method both by single vehicle emission measurements. Similarly it is possible to define an acoustical equivalent for other categories such as motorcycles, buses, etc.

England Standard: Cortn Procedure

The CoRTN procedure (Calculation of Road Traffic Noise) has been developed by the Transport and Road Research Laboratory and the Department of Transport of the United Kingdom in the 1975 and has been modified in the 1988. It estimates the basic noise level L_{10} both on 1h and 18 h reference time. This level is obtained at a reference distance of 10 m from the nearest carriageway edge of a highway.

The parameters involved in this model are: traffic flow and composition, mean speed, gradient of the road and type of road surface. The basic hypotheses of the model are a moderate wind velocity and a dried road surface.

The CoRTN procedure is divided in five steps:

- Divide the road scheme into one or more segments, such that the variation of noise level within the segment is less than 2 dBA,

- Calculate the basic noise level 10 meters away from the nearside carriageway edge for each segment. It depends on the velocity, traffic flow and composition. The traffic is considered as a linear source positioned at 0.5 m from the road surface and at 3.5 m from the carriage edge,

- Evaluate the noise level, for each segment, taking into account the attenuation due to the distance and screening of the source line,

- Adjust the noise level taking into account:

  ○ Reflection due to buildings and facades on the other side of the road and reflective screen behind the reflection point,

  ○ Size of source segment (view angle);

- Join the contributions from all segments to give the predicted noise level at the reception point for the whole road scheme.

Operatively the basic hourly noise level is predicted at a distance of 10 meters from the nearest carriageway, according to the following equation:

$$L_{10}(1h) = 42.2 + 10\,Log(q)\;(dBA)$$

and the basic noise level in terms of total 18-hour flow is:

$$L_{10}(18h) = 29.1 + 10\,Log(q)\;(dBA)$$

where q and Q are the hourly traffic flow (vehicles/hour) and 18-hour flow (vehicles/hour), respectively. Here it is assumed that the basic velocity is v = 75 km/h, the percentage of heavy vehicles is P = 0 and road's gradient is G = 0%. It is also assumed that the source line is 3.5 m from the edge of the road for carriageways separated by less than 5.0 meters.

Subsequently the level will be correct to take into account mean traffic speed, percentage of heavy vehicles and gradient contribute.

In particular the corrections for heavy vehicles and speed are determined using the following expressions:

$$\Delta_{pV} = 33\,Log\left(v + 40 + \frac{500}{v}\right) + 10\,Log\left(1 + \frac{5P}{v}\right) - 68.8\;(dBA)$$

where the mean speed v depends on road type and is reported in equation $L_{10}(1h) = 42.2 + 10\,Log(q)\;(dBA)$ for various roads. The percentage of heavy vehicles is then given by:

$$P = \frac{100\,f}{q} = \frac{100\,F}{Q}$$

where f and F are respectively the hourly and 18- hour flows of heavy vehicles. The value of v to be used in equation $\Delta_{pV} = 33\,Log\left(v + 40 + \dfrac{500}{v}\right) + 10\,Log\left(1 + \dfrac{5P}{v}\right) - 68.8\,(dBA)$ depends on the road gradient. In particular, for roads with gradient, traffic speed in the previous relation will be reduced by the value ΔV which is predicted from:

$$\Delta V = \left[0.73 + \left(2.3 - \frac{1.15\,p}{100}\right)\frac{p}{100}\right]G,\ km/h$$

where G is the gradient expressed as a percentage.

Once the traffic speed is known also the sound level is adjust for the extra noise from traffic on a gradient with the correction $\Delta_G = 0.3G\ \ dBA$.

The noise depends also upon the road surface. In fact for roads which are impervious and where the traffic speed used in expression $\Delta_{pV}\ ...\,(dBA)$ is V > 75Km/h a correction to the basic noise level is applied as:

- $\Delta_{TD} = 10\,Log(90TD + 30) - 20,$ dbA for concrete surfaces;

- $\Delta_{TD} = 10\,Log(20TD + 60) - 20,$ dbA for bituminous surface;

where TD is the texture depth.

If instead v ≤ 75 km/h the corrections are:

- $\Delta_{TD} = -1,$ dBA for impervious bituminous road surfaces;

- $\Delta_{TD} = -3.5,$ dBA for pervious road surfaces.

The model also considers the correction for receiver points located at distances d ≥ 0.4 md from the edge of the nearest carriageway, which is:

$$\Delta_d = -10\,Log\left(\frac{d'}{13.5}\right),\ \ dBA$$

where d' is the shortest distance between the effective source and receiver.

The last correction is the one associated with the propagation obstacles, such as the nature of the ground surface between the edge of carriageway and the receiver point (for example grass land, cultivated fields, etc). or the presence of buildings, walls, barriers, etc.

German Standard: RLS 90 Model

In the Guideline for Noise Protection on Streets, the RLS90 traffic noise model has been defined as an improvement of oldest standard RLS81. RLS90 is an effective calculation model, able to determine the noise rating level of road traffic and, at current day, is the most relevant calculation method used in Germany. The model requires an input of data regarding the average hourly traffic flow, separated into motorcycles, heavy and light vehicles, the average speed for each group, the dimension, geometry and type of the road and of any natural and artificial obstacles.

This model takes also into account the main features which influences the propagation of noise, such as obstacles, vegetation, air absorption, reflections and diffraction. In particular it makes possible to verify the noise reduction produced by barriers and takes into account also the reflections produced by the opposite screens. In addiction this is one of the few models present in literature that is able to evaluate the sound emission of a parking lot.

The starting point of the calculation is an average level L_{mE} measurable at a distance of 25 m from the centre of the road lane. This $L_{mE}{}^{(25)}$ is a function of the amount of vehicles per hour Q and of the percentage of heavy trucks P (weight > 2.8 tons), under idealized conditions (i.e. a speed of 100 km/h, a road gradient below 5% and a special road surface). Analytically $L_{mE}{}^{(25)}$ is given by

$$L_{m,E}^{(25)} = 37.3 + 10 \, Log\left[Q(1+0.082\,P)\right]$$

The next step is to quantify the various deviations from these idealized conditions by means of corrections for the "real speed", the actual road gradient or the actual surface, etc. In particular these correction depends upon whether day (6:00-22:00 h) or night (22:00-6:00 h) is considered. So for each lane the mean level in dBA L_m is calculated as

$$L_m = L_{m,E}^{(25)} + R_{SL} + R_{RS} + R_{RF} + R_E + R_{DA} + R_{GA} + R_{TB}$$

where,

- R_{SL} is a correction for the speed limit;

- R_{RS} is a correction for road surfaces. It's given in a table and depends upon kind of surface and vehicle speed. It ranges from 0 to 6 dB. In particular:

$$R_{RS} = 0.6|g| - 3 \quad for \ |g| > 5\%$$
$$R_{RS} = 0 \quad for \ |g| \le 5\%$$

- R_{RF} is a correction for rises and falls along the streets;

- R_E is a correction for the absorption characteristics of building surfaces;

- R_{DA} is a attenuation's coefficient that takes into account the distance from receiver and the air absorption;

- R_{GA} is a attenuation's coefficient due to ground and atmospheric conditions;

- R_{TB} is a attenuation's coefficient due to topography and buildings dimensions.

In particular the R_{SL} is given by the formula:

$$R_{SL} = L_{Pkw} - 37.3 + 10\ Log\left(\frac{100 + \left(10^{0.1D} - 1\right)P}{100 + 8.23\ P}\right)$$

with

$$L_{Pkw} = 27.7 + 10\ Log\left[1 + \left(0.02v_{Pkw}\right)^3\right]$$

$$L_{Lkw} = 23.1 + 12.5\ Log\left(v_{Pkw}\right)$$

$$D = L_{Lkw} - L_{Pkw}$$

where v_{Pkw} is the speed limit in the range of 30 to 130 km/h for light vehicles and v_{Lkw} is the speed limit in the range of 30 to 80 km/h for heavy vehicles.

Evaluating the $L_{m,E}$ [25] for each lane as described, we can obtain:

$$L_m = 10\ Log\left[10^{0.1L_{m,n}} + 10^{0.1L_{m,f}}\right]$$

where n represent the nearer and f the further lane respectively. Finally the sound pressure level for the street is given by $L_r = L_m + K$.

K is the additional term for the increased effect of traffic light controlled intersections and other intersections.

Italian C.N.R. model

Nowadays the Italian legislation does not suggest any TNM of reference, but the most used by technician is the one developed by the Italian "Consiglio Nazionale delle Ricerche" (CNR) and then improved by Cocchi.

This model represents a modification of the German standard RLS 90, adapted to the Italian framework; a relation between the traffic parameters and the mean sound energy level is supposed and the traffic flow is modeled as a linear source placed in the center of the road. So the equivalent sound level in dBA is given by

$$L_{Aeq} = \alpha + 10\ Log\left(Q_L + \beta Q_P\right) - 10\ Log\left(\frac{d}{d_0}\right) + \Delta L_V + \Delta L_F + \Delta L_B + \Delta L_S + \Delta L_G + \Delta L_{VB}$$

where Q_L and Q_P are the traffic flow in one hour, related to light and heavy vehicles respectively, d_0 is a reference distance of 25 meter and d the distance between the lane center and observation point on the road's edge. Then:

- ΔL_V is the correction due to mean flux velocity defined in the following table;

Flux mean speed (Km/h)	ΔL_v (dBA)
30-50	+0
60	+1
70	+2
80	+3
100	+4

- ΔL_F and ΔL_B are the correction for the presence of reflective façade near the observation point (+2.5 dBA) or in opposite direction (+1.5 dBA) respectively.

- ΔL_S is the correction for the road's pavement defined in the following table;

Road's pavement	ΔL_S (dBA)
Smooth Asphalt	– 0.5
Rough Asphalt	0
Cement	+1.5
Rough pavement	+4

- ΔL_G is the correction for a road's gradient greater than 5%. The correction value is +0.6 dBA for each % gradient over 5%.

- ΔL_{VB} is a coefficient that takes into account the presence of traffic lights (+1.0 dBA) or slow traffic (-1.5 dBA).

Whilst all the cited parameters have a general validity, independent by countries (because related just to physical or urban parameters), the α e β parameters are influenced by characteristics of countries roads and vehicles. In particular α is related to noise emission from the single vehicles and β is the weighting factor that takes into account the greater emission of heavy vehicles (very frequently for Italian roads α = 35.1 dBA and β = 6 are assumed).

French Model: NMPB Routes

The European directive 2002/49/CE for what concern the traffic noise prevision model suggests to use the official interim French standard model "Nouvelle Methode de Prevision de Bruit" or simply NMPB-Routes-96.

This method has been developed by different French Institutes of Ministère de l'Equipement (CSTB, SETRA, LCPC, LRPC) and represents an improvement of an oldest one defined in the "Guide de Bruit" of 1980, that takes into account the meteorological conditions and the long distance (d > 250m) prevision, as suggested in the ISO 9613.

Nowadays it represents one of the most used TNM, being also integrated in some commercial software such as CadnaA™ by 01dB.

In the 2000, under request of SETRA, a revision of NMPB-Routes-96 started, bringing to the NMPBRoutes-2008.

The method is based on the concept of propagation path. Several paths between a source and a receiver can exist, depending on topography and obstacles and, at each of them, a long term sound level $L_{Ai,LT}$ may be associated.

Despite of previous models, NMPB takes into account the standard meteorological conditions, as suggested by the ISO 9613, to adjust the prevision on long-period. They are classified in two types: meteorological conditions "favorable to the propagation" and "homogeneous acoustical conditions" (corresponding to the conditions used in the oldest French model).

So, the long-period prediction level for each path $L_{Ai,LT}$ is evaluated adding the terms corresponding to this two conditions:

$$L_{Ai,LT} = 10 Log\left(p_i 10^{\left(0.1 L_{Ai,F}\right)} + \left(1 - P_i\right) 10^{\left(0.1 L_{Ai,H}\right)} \right)$$

where $L_{Ai,F}$ and $L_{Ai,H}$ are the global levels evaluated respectively for favorable and homogeneous conditions and p_i represent the probability of occurrence of favorable conditions.

These levels are calculated for each octave band and for each path from the source, according to the following formulas:

$$L_{Ai,F} = L_{A,w} - A_{div} - A_{atm} - A_{grd,F} - A_{diff,F}$$
$$L_{Ai,H} = L_{A,w} - A_{div} - A_{atm} - A_{grd,H} - A_{diff,H}$$

For each path the algorithm computes three different attenuations: the geometrical spreading A_{div} and the atmospheric absorption A_{atm}, that are the same in both formulas, and the boundary attenuations A_{bnd}, which depends on the propagation conditions and are determined by ground effect (A_{grd}) and diffraction (A_{diff}).

The sound power level, $L_{A,w}$, is evaluated considering the hourly flux Q, reported in equation $L_{m,E}^{(25)} = 37.3 + 10\, Log\left[Q(1+0.082\,P)\right]$ and directly obtaining the equivalent hourly level in dB(A), E, associated to a single light or heavy vehicle. By this procedure the pointlike source acoustical power representing the road is given by:

$$L_{Awi} = \left[\left(E_L + 10 Log\, Q_L\right) + \left(E_p + 10 Log\, Q_p\right)\right] + 20 + 10 Log\left(I_i\right) + R(j)$$

where E_L and E_p are the emission levels obtained from equation $L_{m,E}^{(25)} = 37.3 + 10\, Log$ $\left[Q(1+0.082\,P)\right]$ for light and heavy vehicle, I_i the length in meter of considered road and R(j) is the value of normalized noise spectra from CEN 1793-3 that take into account the frequency behavior of propagation.

The predictions of NMPB-Routes-96 have been validated on a great number of experimental campaigns with various topography and meteorological conditions, founding a very good agreement with the noise data but generally an overestimate level is found in downward propagation conditions. That's why SETRA required the revision of the model. The NMPB-Routes-2008 presents a better estimation of noise level in downward condition, takes into account reflections on embankments, is able to evaluate the correction due to diffraction by low barriers and has implemented other minor corrections.

Railway Noise and Measurement

The Sources Railway Vehicles Noise

When the movement of trains along the track creates sound waves that result from:

- Interaction between wheel and rail,

- Interaction of the surfaces in contact in circuits of wagon,

- Interaction of the outer surface of locomotives and wagons with air in motion,

- Interaction between the track and the substrate on which is placed,

- Interaction between wheel and brake,

- Operation of basic and additional equipment in locomotives and wagons.

It is common that in the circumstances under which it accrues, railway vehicles noise is classified in the following categories:

- Rolling noise,

- Wheel squeal,

- Curve squeal,

- Aerodynamic noise,

- Bridge noise,

- Ground noise and vibration

- Internal noise and vibration,

- Other source of railway noise.

Based on the frequency spectrum, the sources of railway vehicles noise are classified according to the following table:

Table: Frequency range for different types of railway noise.

Noise type	Frequency range [Hz]
Rolling	30 - 5000
Flat spots	50 - 250 (as a function of speed)
Ground borne vibrations	4 - 80
Structure borne noise	30 - 200
Top of rail squeal	1000 - 5000
Flanging noise	5000 - 10000

Individual impact of the sources of noise on the overall noise level of trains in motion depends on many factors such:

- Weighing railway vehicles,

- Radius of curve,

- Speed,

- Number of axles per vehicle,

- Number of vehicles in the train

- Type of rail,

- Type of brakes and brake-block material,

- There is no contact lubrication,

- Method of modification friction.

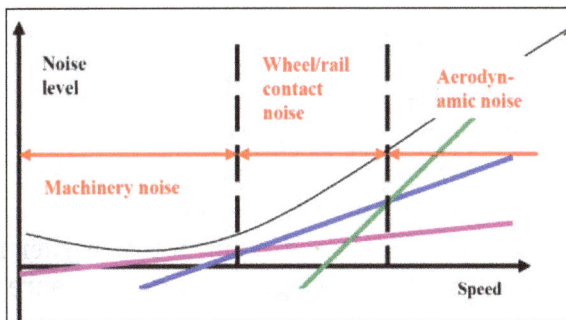

The noise level comprises contribution from the machinery,
the wheel/rail contact and from aerodynamic flow.

Some of these factors are primary factor for the occurrence of certain types of noise, while on other less common sources of noise. So, for example, the onset of squeal (regardless of whether the wheel or curve), radius of curve has a dominant influence. Speed of movement is the primary factor in many sources of noise and it was sometimes the main factor determining the source of noise that will be dominant.

Form that calculates the noise during acceleration of the train, in function of the speed the train is:

$$L_{A,max} = 10 \log (10^6 \cdot K + 447 \cdot V^3) \, [dB]$$

where,

$L_{A,max}$ – maximum A-weighted sound pressure level recorded by a sound level meter set to the standard "Fast" response over the time for the train to pass the microphone.

K – traction noise factor in accordance with table below.

V – train speed in km/h.

Table: Noise factor K for traction power ratings.

Vehicle Power Range [MW]	Diesel traction	Electric traction
Above 1.0	3160 (95)	1000 (90)
0.3 to 1.0	1260 (91)	400 (86)
Below 0.3	500 (87)	160 (82)

The noise emitted from a train when running at is maximum normal operating speed, with traction equipment idling gets the form:

$$L_{A,eq} = 40 + 20 \log V + 10 \log N \cdot 10 \log t \, [dB]$$

where,

N – number of axels in train.

t – time for train to pass microphone [s].

Otherwise, these values of noise levels and noise levels of the value of static (immovable) of the train, obtained by measuring the vertical range of the microphones on the side of the train, the horizontal distance of 7.5m from track centerline at heights of 1.5m to 3.5 above the top of the rails. If we look at the value of C-weighted sound pressure level, the values obtained must not be higher for more than 7 dB compared to those obtained via the A-weighted sound pressure level.

Rolling Noise

Rolling noise is the result of vibration wheel and vibration tracks that appear in direct contact with the wheel/rail due to roughness (asperity) on the surfaces in contact. It is dominant noise source at speeds of conventional trains.

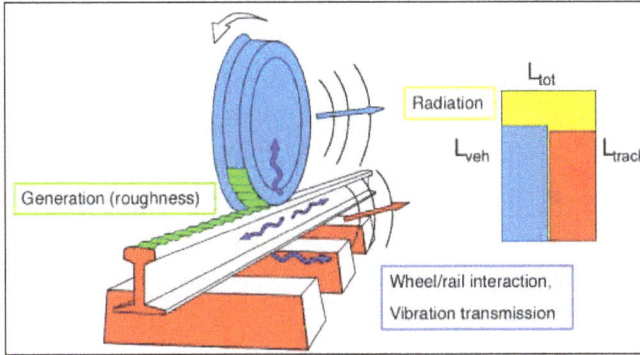

Illustration of the mechanism of generation of the rolling noise.

Rail roughness occurs due to prolonged use and the passage of large number of train compositions, influenced by weather conditions, especially in winter due to freezing of water in micro cracks, due to the rail wheel skid etc. Roughness point increase in longterm use of wheel traffic, but also depends on the type of brakes that are fitted on the wagons, the entire state assembly wheel set or bogie.

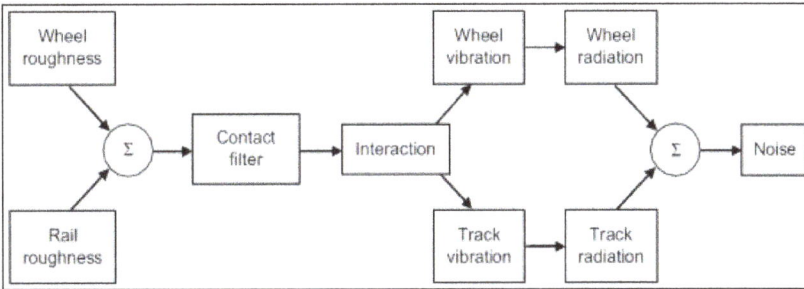

Model for rolling noise generation.

It is difficult to accurately determine whether the noise comes from the wheel or the infrastructure, although it is known that a high level of noise comes by irregularity of wheels. Irregularities on wheel form waves on the surface of wheel the wavelength values of 5- 500mm. If the λ [m] denote the wavelengths and the v [m/s] speed, we can calculate the frequency of the waves through the following relationship:

$$f = \frac{v}{\lambda}[Hz]$$

When contact the wheel/rail, contact isn't in the spot, but when contact the surface of irregular shape, the effects of roughness in the contact zone are small. They come to the fore the occurrence of wave frequencies from 1 to 1.5 Hz and at speeds exceeding 160 km/h or at lower frequency waves at lower speeds.

Therefore, the A-weighted sound pressure level is usually calculated as proportional to the logarithm of speed:

$$L_p = L_{p0} + N \cdot \log_{10}\left(\frac{v}{v_0}\right)$$

where:

Lp_0 – the sound level at a reference speed V_0, N – values of the speed "exponent", determined from measurements on the basis of linear regression, are usually found to be between about 25 and 35, with a typical value of 30.

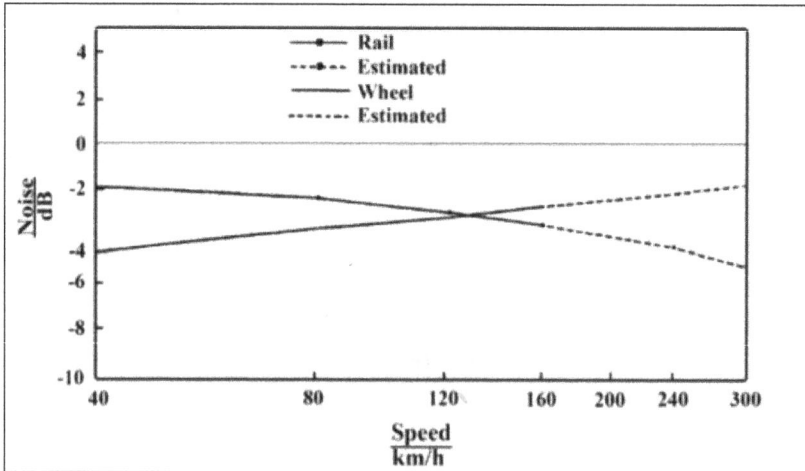

Estimated components of wheel and rail noise relative to
A-weighted level as a function of train speed.

These considerations suggest that doubling the speed corresponding increase A-weighted level of 8 to 10 dB. Speed affects the dynamics behavior of the wheel and the dynamics behavior of the rail, but also on their interaction and must be taken into account when considering the problem.

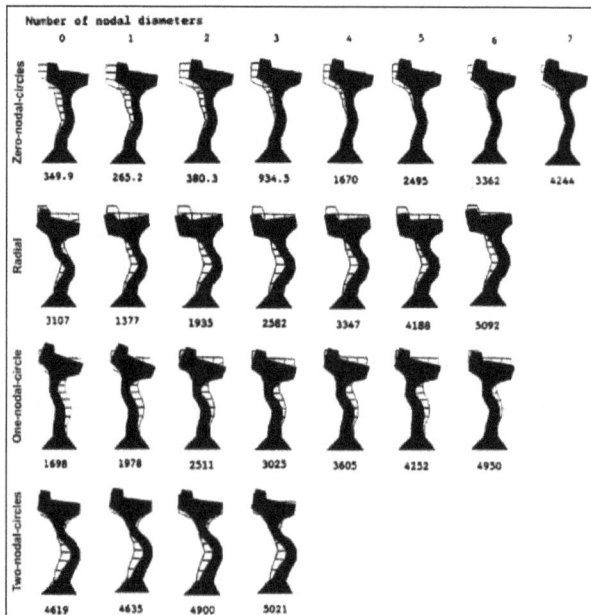

Modes of vibration of a UIC 920mm standard freight wheel
shown in cross-section and natural frequencies in Hz.

Dynamics behavior of the wheel is seen through the analysis of vibration modes and natural frequencies. Railway wheel design is slightly damped resonant structure. Due to the movement and interaction with the rail, the wheel acts like any other element that vibrate own and natural oscillations. The wheel is like new axle-symmetrical and when is new has no impact places (noodle place diameters) while in use to a greater or lesser extent they appear. Because of this phenomenon comes to the valuation of such oscillations is shown for a wheel diameter of 920-mm for freight wagons in cross-section.

Especially for mid to wheel and rail is one mechanical system, due to mutual contact wheel oscillates near the rails or track. Rail is connected to a rigid threshold which can be wood or concrete, that all relies on a set of clay and partly covered with ballast. If we accept that link of the rail with the threshold is absolute rigid and the threshold motionless, there was a rail oscillations due to vertical forces that occur in contact with the wheel/rail. Mostly it comes to frequency of the order of 500-1000 Hz.

Wheel Squeal

The main cause of wheel squeal is curving noise. It consists of two kinds of squeal and:

- Squeal of flange of wheel,
- Squeal in contact the wheel/rail due to the appearance longitudinal stick-slip and lateral slip.

Longitudinal stick-slip phenomenon is the noise of high frequency noise due to the different speeds of the wheels of wheel set on inner and outer rails. The radius of curve, the model (geometry) of wheel and the profile of rail, as well as speed, are the main factors affecting the level of noise that occurs in these cases. The radii of curve up to 600m can reduce this noise compensation speed, while for smaller radii can reduce noise by grinding rails.

Lateral slip is the noise that occurs due to lateral slip surface of wheel on the upper surface of rail, and that is the main cause of this type of noise. This type of noise is most evident when entering the curve with a cant deficiency of rail.

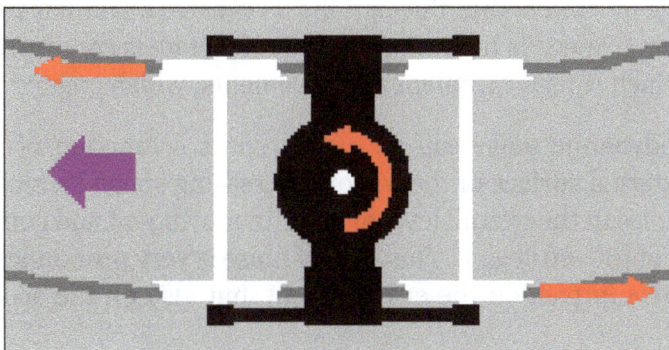

Schematic of a Rail bogie in a curve.

On that occasion, it creates noise wreath wheel while touching the inside of the rails. By the effect of centrifugal force of the internal crown, wheel moves away from the inner (or lower) rails, while the crown wheel on the outside (or higher) rail contacts the inner edge of the son. Profile of wheel should be allowed to touch the surfaces in such cases less due to increased wear. Reduction of friction lubricants are often used. Frequency band noise resulting from wheel squeal is in the range 1400-1600 Hz, depending on the geometry of wheel and rail.

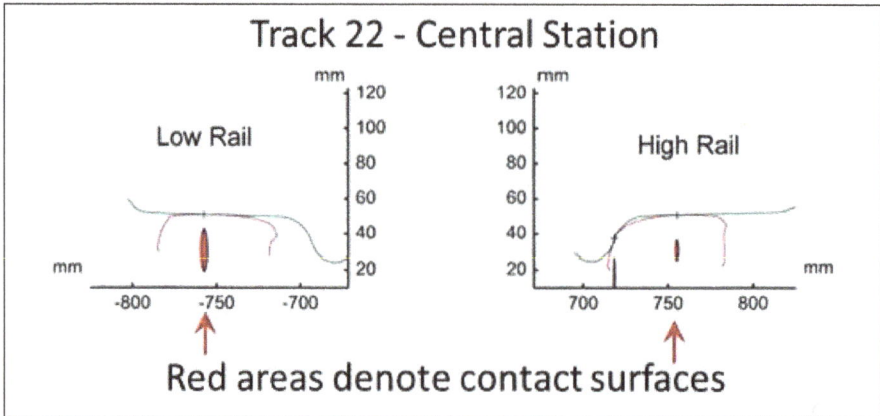

Actual wheel and rail profiles.

Frequency range of noise known as wheel squeal is around 3000-4000Hz, and these sounds are very high pitched and unpleasant for a person's hearing. Wheel squeal occurs in freight wagons because of the high pressure shaft, which causes an increase in wheel and rail surfaces in contact. There are several solutions to reduce these types of noise such as the Japanese railways employ a method of partial lubrication oil spray special rails that are on the front of the locomotives near the top of rails, dropping oil only when entering a corner to reduce wheel squeal. Australian railways uses specially designed blockers that mounted the suspension to reduce wheel slip by rail.

Aerodynamic Noise

Due to the passage of solid bodies through the air creates a laminar flow of air around the body. However, if speed is high, close to the solid state is coming to the emergence under pressure which causes turbulent air movements, which created sound waves.

The level of aerodynamic noise can be expressed as a function of train speed and strength of the external surface of the wagon. Increasing speed increases the influence of aerodynamic noise in the overall level of noise in movement and composition is typically in the range of $(60\text{-}80)\log_{10} V$. This type of noise is very pronounced in high-speed train and a negative impact on the environment, but also on the acoustic comfort of travel.

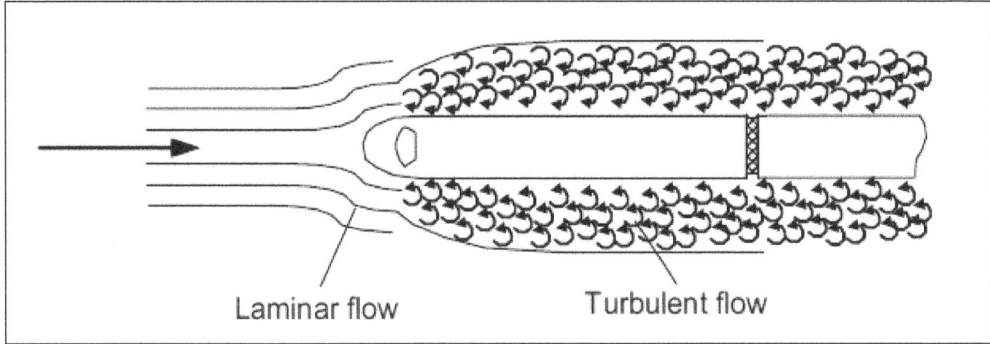

Aerodynamic moving.

The aerodynamic noise generated on the front of the locomotive, in the spaces between the wagons, carriages on the sides if the sides are made of elastic material, the burden is not compressed during transport open types of freight wagons on the sides due to gust, the pantograph, etc. The values of aerodynamic noise in some places are different, and approximates the sound energy is expressed that occurs:

$$W_{rad} \propto \frac{\rho_o \cdot U^8 \cdot l^2}{c_0^5}$$

where:

l – the width of the flow,

U - flow velocity,

c_0 – the speed of sound,

ρ_0 – the fluid density.

Measurement of aerodynamic noise carries out a number of microphones arranged in height and length in the "stellar" form, or in concentric circles.

The measurement was shown that the size of the aerodynamic noise at high-speed train TGV (when the velocity of 200 km/h) ranges from 92dB on the frontal glass and the rear locomotive pantograph up to 79dB on the bogie. At high-speed train ICE these values are similar values. Acoustic noise, caused by the movement of railway vehicles, a low-frequency sound waves with noise levels of 60 dB for a speed 100 km/h to 110 dB at speeds of nearly 500km/h.

Reduction of aerodynamics noise is done by defining the appropriate aerodynamic shape of locomotives and trains, but mounting the protector of the basic elements of aerodynamic pantograph. Spaces between the wagons to shut special elastic materials that prevent the occurrence of turbulence in the spaces between the wagons, and again the composition do not violate the security of enrolling in a corner. It also made the side "cover" bogie which prevents air streamlines that pass through a bogie. An interesting choice of engineers high-speed train Eurostar's to get the budget to open diameter of 75mm and a depth of 0.5m makes reduction of frequencies over 300Hz to about 10dB.

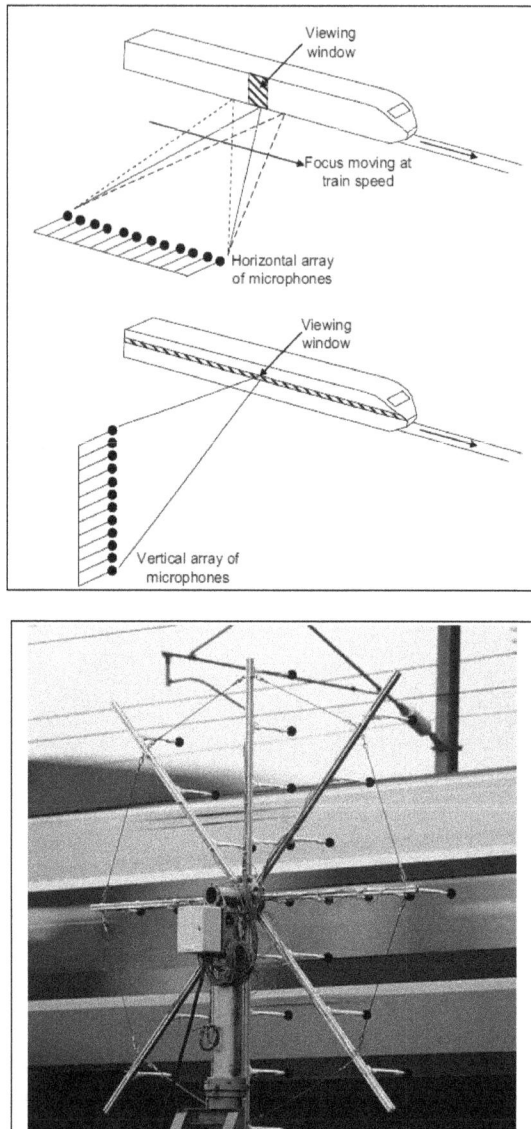

Principles of microphone array: Horizontal array with swept focus, vertical array and example of star-shaped microphone array consisting of 29 microphones.

Ground Noise and Vibration

Besides the noise, it is expressed due to the passage of vibration phenomena of the composition. Specifically, the passage of the train movement can be seen as a rigid body on elastic base. Elastic properties of the substrate condition are creating waves in the environment that we feel as low-frequency vibrations of the ground. This phenomenon is most evident when cargo composition moves with a large number of wagons.

The movement of waves through the surface is expressed in all directions:

- Along the direction of movement,

- Sideways movement and direction and

- Perpendicular to the direction of movement.

There are two methods of calculating the level of vibration when going composition:

Vibration dose values (VDV) – Defined to quantify intermittent vibration. The VDV for a single event is:

$$VDV = \left[\int_0^T a^4(t)\,dt \right]^{0.25} \left[m / s_{1.75} \right]$$

where:

T – the duration of the event,

$a(t)$ – the frequency weighted (filtered) acceleration as a function of time.

The total values of train are:

$$VDV_T = \left[VDV_1^4 + VDV_2^4 + VDV_3^4 + ... \right]^{0.25}$$

KB value – Uses a running root-meansquare vibration velocity measurement (based on a 0.125 second time constant).

$$KB_{FTr} = \sqrt{\frac{1}{T_r} \sum_j T_{e,j} \cdot KB_{FTm,j}^2}$$

where:

T_r – the evaluation period,

$T_{e,j}$ – the exposure period of each event j,

$KB_{FTm,j}^2$ – the average of the maximum filtered r.m.s. signal values during each 30 sec interval of the whole event.

If there is the substrate steel construction of the bridge, then we talk about the bridge noise. It is specific due to increased enforcement and self-oscillation. The uniqueness of the bridge noise is seen in the fact that there is partial ability to reduce the noise and the use of the fastener stiffness, rail damping, and ballast mats, damping of bridge structure, plate thickness, barriers and enclosures etc.

Aircraft Noise

Sources

There has been a substantial increase in the noise nuisance from subsonic aircrafts over the last two decades. There are several reasons for this. In the first place, noisier jet-engined planes have progressively replaced the earlier piston-engined and turbo- prop types. The latter types of planes, although still noisy, were relatively quiter in comparison with jet-engined aircraft.

Secondly, there has been a steep rise in the number of civil aircraft. This has resulted in a considerable increase in the number of air movements. When compared to developed countries of Europe and America, the number of civil and military airports in developing nations is not very large.

The main noise disturbance due to aircraft is confined to a radius of about 16 km around the airports. But there are many people who work or live under the flight paths connecting air ports. For these people, the noise from the aircraft passing overhead is inescapable.

Aircraft noise is variable and intermittent. It is not continuous as in the case of road traffic noise. There are peak noise levels when aircrafts are flying overhead, or are taking-off and landing at the airports. The peak frequency varies with the number and the types of aircraft, and the operational height.

Sources of Aircraft Noise

The chief producers of noise in the aircraft are various components of the propulsive system. These include the engine, propeller or rotor, jet engine compressor, turbines and jet flow, etc. The noise generated by the motion of the aircraft through the air is confined to boundary layer noise which is really an aerodynamic phenomenon.

This, in turn, is mainly responsible for the secondary noise generated by the skin vibration, observed particularly inside the aircraft in flight.

The contribution of this noise to an outside observer is almost entirely masked by the propulsion noise. Other pure sources of aerodynamic noise are the static pressure fields moving with very low flying aircraft, and shock waves and sonic booms generated by supersonic flight. These two (i.e., static pressure fields and sonic booms) are separate phenomena.

Sound Pressure Levels of Aircraft

Typical overall sound pressure levels are generated by four distinct types of aircraft.

These are:

- Propeller driven aircraft,

- Jet-engined aircraft (for transport),

- Jet fighter,

- Jet bomber.

Each of these four types may undertake three different climb out positions, viz:

- Steep climb,

- Normal climb, and

- Shallow climb.

There are twelve possibilities for the typical noise levels explained in the diagram. These diagrams would establish the sound pressure levels for aircraft in level flight, this time as functions of slant range instead of the distance from take-off.

But neither of these two types of sound pressure level measurements conveys any specific information about the sound "footprint" (actual distribution of sound on either side of the flight path). This can only be given by actual sound pressure contours plotted with the help of measurements taken at a large number of different points.

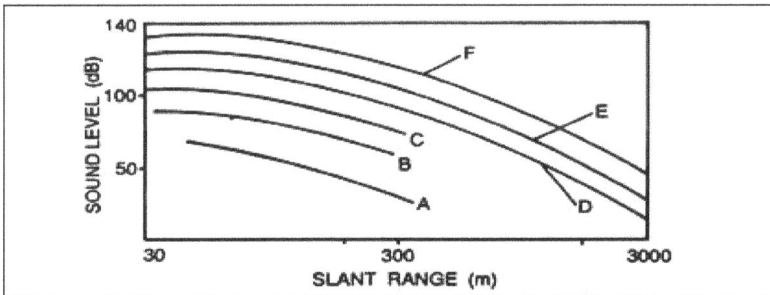

Sound pressure level for various type f aircraft in level flight, as a function of slant height (A-small light aircraft: B-Single engined light aircraft: C- double engined light aircraft; D-four engined aircraft; E-jet fighter; F-large military jet).

Piston-engined Aircraft

The main sources of noise with the piston-engined aircraft are the engines and propellers. Among these, the propeller noise is often predominant except at fairly low tip speeds and low power levels.

The propeller noise consists of two components:

- Discrete tones at the blade speed frequency and its harmonics,

- Vortex noise or broad-band noise which predominates above about 1000 Hz. At very low tip speeds, the vortex noise can exceed the noise due to rotation of the propeller at all frequencies. Mathematical analysis of these two components of propeller noise is possible. It becomes complex, however, if forward speed and blade load distribution are taken into account.

The sound field for the vortex field generation is directive. It is characterized by a four-lobe pattern. In this diagram, the arrow indicates the direction of flight of the aircraft. The sound pressure level under the angle of maximum radiation is usually of the order of 4 dB higher than the space average.

This angle (of maximum sound radiation) is roughly 25-30° inclined backwards from the plane of propellers.

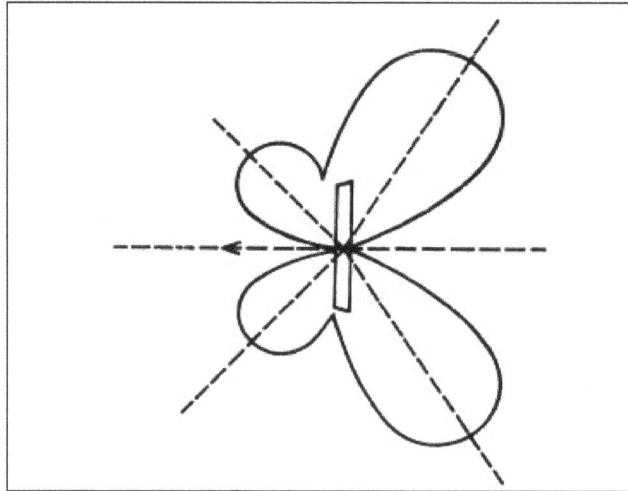

Four-lobed sound field pattern for the vortex field generation.

The vortex noise level can be calculated with a reasonable degree of accuracy (+ 10 dB) from the empirical relation

$Lp = 10 \log_{10} [10^{16} \, k \, A_b (Vo._7)^6]$ dB,

where Lp = Vortex noise level in dB,

Ab = total area of propeller blades (in m²),

Vo.7 = velocity of blade element at 0.7 radius (in m/s),

k = an empirical constant

Similarly, peak frequency f_{max} of the vortex noise can be obtained from the following formula:

$$f_{max} = 0.13 \frac{Vh}{0.7} Hz.$$

where V_h = helical tip speed of the propeller (in m/s),

R = radius of the propeller (in m).

The typical contribution of vortex noise to the overall noise of a piston which refers to a specific diameter of the propeller and specific input horsepower.

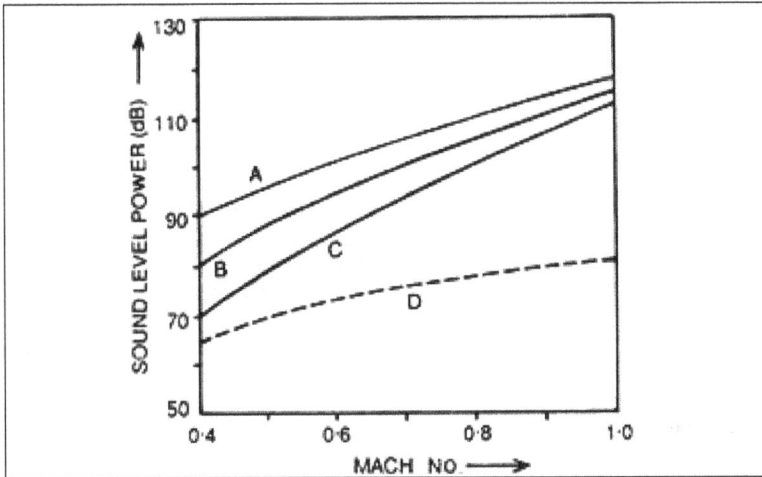

Sound power level for 1.000 horse-power aircraft with 30 cm diameter propeller, as a function of Mach no.
(A-propeller with 3 blades; B-propeller with 4 blades; C- propeller with 5 blades; D- Vortex noise level).

Engine Noise in an Aircraft

The engine noise in an aircraft is derived almost entirely from the pulsating exhaust flow. Its fundamental frequency is, therefore, equal to the average cylinder firing frequency of the engine. For the direct analysis of sound output in this case, the engine is treated as a simple source. But this approximation does not hold good for harmonics.

In addition to equation $L_p = 10 \log 10 [1016 k Ab(Vo.7)6]$ dB given earlier, the overall sound power level L_p can also be estimated directly from the following empirical relation:

$L_p = 125 + 10 \log_{10} W_t$ dB (ref 10^{-13} watts),

where W_t = total horsepower delivered by the engine.

By treating the engine as a simple point source, the sound pressure level L_r at a distance r is given by

$L_r = L_p - 10 \log_{10} 4 T\pi/r^2$ (ref 0.0002 microbars).

For low powers, in general, there is a typical decrease in the sound pressure level for each successive octave of 2.5-3.0 dB, with a leveling off at higher frequencies. For engines operating at maximum power (particularly large engines), the power level is substantially the same in all octave bands, at a level of about 7-8 dB below the overall level.

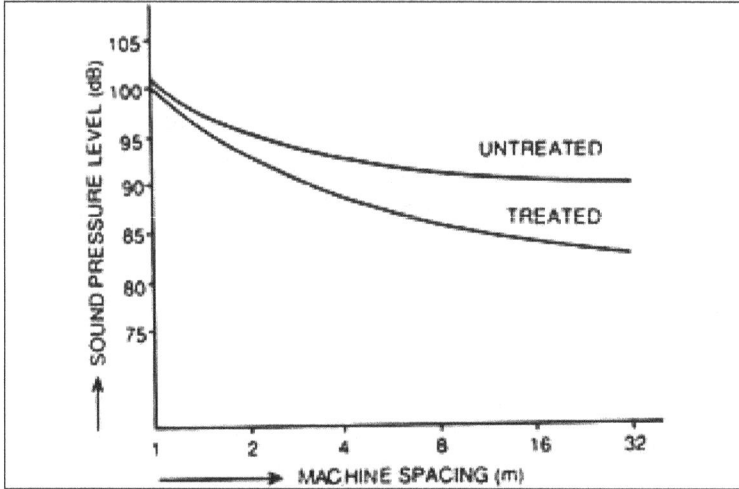

Sound power level for 1,000 horse – power aircraft with 30 cm diameter propeller, as a function of Mach no
(A-propeller with 3 blades; B-propeller with 4 blades; C-propeller with 5 blades; D-vortex noise level).

Reduction of Engine Noise

The standard method employed to reduce exhaust noise is silencing. The application of silencers is, however, limited to straight-through type silencers to minimise backpressure. Moreover, the use of silencers is also likely to modify the sound spectrum of the engine noise.

It may suppress some frequencies and amplify others, depending on the relative phases of the individual exhaust pulses as they meet the exhaust system.

In addition to exhaust noise, the engine may also generate noise by engine-excited vibrations. These can be transmitted through the engine mounts to the airframe. This, however, is only likely to affect noise levels inside the aircraft.

This type of noise is generally negligible in the aircraft of modern design where adequate attention is given to isolation of the engines at the point of mounting. In older types of aircraft, however, engine vibrations may be responsible for vibrational waves being set up in the fuselage.

The frequencies of these vibrations are, in general, below 600 Hz and they decrease with increasing distance from the source. The usual treatment of such engine-ex- cited vibrations was to stiffen the forepart of the fuselage (on single-engine aircraft) or the region of the plane of the propeller (on multi-engine aircraft).

Modern treatment for the engine-excited vibrations, however, is to suspend aircraft engines on rubber-in-shear mounts so as to decouple the six modes of translational and rotational vibrations, with a natural frequency for the suspension of less than 50 Hz. For all vibrations above this natural frequency, attenuation increases by 6 dB per octave.

Jet-prop Aircraft

In this type of aircraft, the propeller is driven by a turbojet instead of an internal combustion engine. When this is the case, the jet exhaust contributes only about 10% of the total thrust and is thus working under comparatively low power level conditions.

As a result, jet noise is usually far less significant than propeller noise. In spite of this, it may appreciably modify the overall noise spectrum by "filling in" between any discrete frequencies which may be present.

Jet-prop aircraft is appreciably quiter than the combination of a propeller and a piston engine for equal thrust. This difference in the noise levels of two types of aircraft is of the order of 10 db. Under particular conditions, and particularly at low thrust levels, compressor noise may predominate.

It is easily identified as a high pitched whine due to fan noise radiated from the engine intake. This noise is most noticeable to an observer positioned in front of the engine. To an observer on the ground, therefore, compressor whine is a noticeable characteristic of a jet-prop aircraft approaching to land.

Sonic Boom

The phenomenon of sonic boom occurs due to the shock waves produced in supersonic flight, i.e., flight of aircraft at speeds larger than the speed of sound in the air. In supersonic flight, the shock waves generated in the near field to the aircraft are of a complex, multi-peaked pattern.

These develop into a more clearly defined double-peak pattern in the intermediate field and finally into the well-known "N-wave" pattern in the far field.

The N-wave shock pattern with its separate bow and tail shocks is the "pressure signature" of the aircraft in supersonic flight. It can be expected to sweep over the ground at the same velocity as the aircraft and is responsible for the sonic boom and associated phenomena.

The separation between the two waves (bow and tail shocks) is determined by the length of the aircraft. For a large aircraft, two separate booms may be heard.

For the smaller aircraft, on the other hand, the two waves may be sufficiently close together for the ear to hear a single boom, although there are still two distinct pressure pulses. Moreover, a larger aircraft will not only extend the time duration of the N-wave but also increase the over-pressures.

Usually the duration of the ground pressure wave for a small fighter aircraft is of the order of 0.1 seconds. The same for a supersonic bomber may be around 0.3 seconds. The

SST (supersonic transport) aircraft, on the other hand, may produce a sound pressure wave of 0.3 to 1.0 second duration, depending on their length.

Besides length, the duration of the pressure wave is also modified by the altitude of the flight path of the aircraft relative to the ground. Although the characteristic N-wave still reaches the ground, any departure from the horizontal flight will shorten the distance between the low and tail pressure peaks and, consequently, the duration of the pressure pulse.

Instead of the idealized linear form both bow and tail shocks diverge with distance. As a result, the far field "pressure signature" may depart to some extent from the standard N-wave configuration.

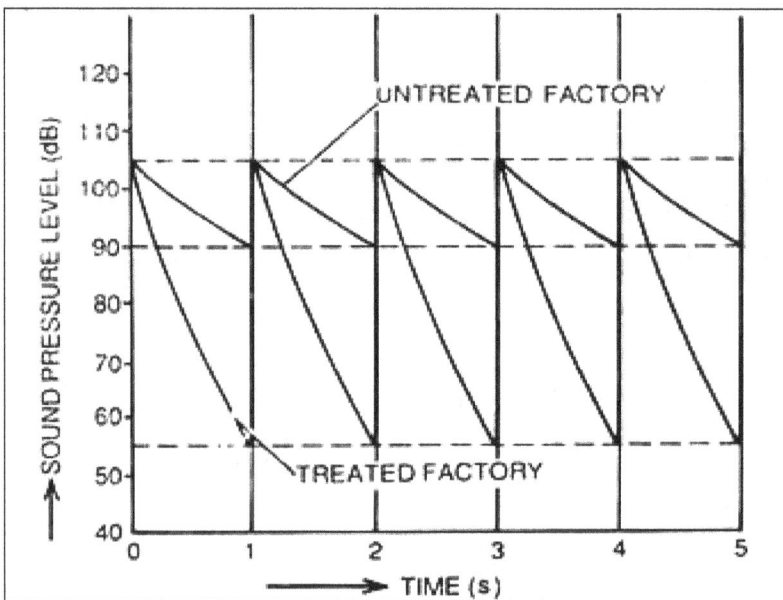

Airport Noise

Noise nuisance near an airport is usually characterized by the "Noise and Number Index" (NNI). Contours based on the NNI are now generally accepted as providing a convenient and relatively straightforward method of assessing the annoyance caused by aircraft operating in the vicinity of an airport.

The Noise and Number Index (NNI) of a point on the ground is defined by,

NNI= L_{ap} + 15 log 10 N-80

Where L_{ap}= average peak noise level of aircraft,

N = number of flights per day.

The derivation of this index makes it necessary to perform the computation for an average 12-hour day in the busy period of the year.

The pattern of noise exposure of the area around an airport can be obtained, therefore, if the following information is available:

- The flight paths of arriving and departing aircraft,

- The number and types of aircraft on those flight paths (for the prescribed period of calculation) and

- The noise level emitted by those particular types of aircraft.

Importance of Noise and Number Index (NNI)

The NNI is particularly valuable for assessing the potential noise nuisance likely to result from future developments and extensions of the existing airports or from the construction of new airports. When the NNI is used for this purpose, the effects of changes in various parameters can be assessed simply by changing the inputs to the calculation.

NNI, is therefore, a very good diagnostic, as well as predictive, method of noise assessment.

Subjective reaction to the NNI, as determined by objective measurement, can be summarized as follows:

- In areas of NNI greater than 35, aircraft noise begins to become a significant reason for discontent with living conditions. In areas where NNI is greater than 55, aircraft noise can be considered as intolerable.

- In areas between 40-60 NNI, no major residential development should be allowed, but in fill development may be allowed subject to adequate sound-proofing being incorporated into the design of the dwellings.

- In areas of greater than 60 NNI, no major residential development should be allowed.

This grading of airport noise nuisance against NNI should be considered carefully in the context of planning for residential development. It offers guidance to a planning authority when considering applications for such development.

Ground Running Aircraft Noise

Ground running is an essential feature of airport operation. This is especially true for airports where maintenance is carried out. To some extent, the introduction of jet aircraft has eliminated much of the noise associated with ground running prior to take off.

From the view-point of noise nuisance, a nice feature of jet aircraft is that they do not need lengthy running up period necessary for aircraft with piston engines for the cylinders to reach their correct operating temperature.

But jet aircraft do have to be ground run to test them after maintenance work. When this happens, they are much noisier than piston-engined aircraft. Ground running of jet engines requires attenuation of the order of 20-30 dB to achieve satisfactory noise suppression.

The arrangement for this purpose (noise suppression) should ideally consist of an exhaust muffler and an intake silencer. Both of these should be fitted with an acoustically tight seal tailored to fit the concerned aircraft.

When this is not possible, the other alternative is to do the ground running in a specially designed test cell (for individual engines removed from the aircraft) or in an acoustic run-up hanger (for complete aircraft). Such systems, however, involve highly elaborate constructions and considerable cost. The employment of such systems, therefore, is largely confined to major aircraft maintenance bases and development areas.

Calculation Method

This topic describes a new numerical method for calculating sound exposure levels at any ground locations resulting from operations of jet and propeller driven aircraft in the vicinity of an airport. The procedures assume that reference noise and performance data are available for each aircraft involved. The new fundamental element of the procedures is a method for calculating the A-weighted sound exposure levels (SEL) that would be produced, on average, by any specific aircraft when performing any specified operation. Procedures are given for calculating sound exposure levels for individual aircraft operations and for the average sound levels produced by the cumulative effects of a series of different aircraft operations. The principal purpose of using these numerical methods which calculate contours of equal average sound level is to assist in land use planning around airports.

Performance Calculation

The aircraft used in this simulation has a maximum takeoff weight of 32000 Kg and it is powered by two low bypass jet engines.

Initial climb speed:

The calibrated initial climb speed is:

$$V_c = C\sqrt{W}$$

where, $C = 0.574$, $V_c = 0.574\sqrt{71000} = 152.9$ Knots.

Equivalent takeoff ground roll:

The equivalent takeoff ground roll distance is:

$$s_g = B\theta_{am}\left(W / \delta_{am}\right)^2 / \left[N P_n / \delta_{am}\right]$$

where, B = 0.0149 and N = 2 (Number of engines).

Therefore:

$$S_g \left\langle \frac{1..0(0.0149)(71000)^2}{2(8520)} \right\rangle = 4408 \; ft$$

Initial climb (Sea Level to 1000 ft):

The average geometric climb is:

$$S_g \frac{1000}{\tan \gamma} = \frac{1000}{0.1654} 6046 \; ft$$

Acceleration and flap retraction:

The horizontal, or ground track, distance, is:

$$S_g (1/2g)(0.95)\left(V_{tb}^2 - V_{ta}^2\right)/\left\{\frac{N/F_n/\partial_{am}}{(W/\partial_{am})} - R_{avg} - \frac{V_{tg}}{101...3V_{tavg}}\right\}$$

Continued climb:

Average altitude = 2130 ft:

$$\Delta h = 3000 - 1259 = 1741 \;\; ft$$

$$S_c = \frac{\Delta h}{\tan} = \frac{1755}{\tan 7.13} = 13906 \; ft$$

Noise Calculation

The representative aircraft has noise-power-distance data relating to SEL power setting and minimum slant distance described in table. The calculation of SEL for any point P on the ground can be now accomplished.

Representative Locations:

Consider three locations which are 610 m (2000 ft) to the side of the takeoff ground track.

The three points are:

P(1): (-1500,2000), P(2): (1500,2000), P(3): (7000,2000), Start of roll: (0,0)

Sound exposure level at point P1:

The sound exposure level at P(1) is dependent upon the aircraft to observer distance at the start of roll, the directivity angle, and the takeoff power. The equation for the noise calculation is:

$$L_{ae}(P_1) = L_{ae}(P, d) + \Delta v - \lambda(0, r) + \lambda_1$$

The speed adjustment for duration is:

$$\Delta v = 10\lg(160/32) = 7.0 \ dB$$

Lateral attenuation is computed as described in reference:

$$\lambda(0,r) = 15.09\left(1 - e^{-2.088}\right) = 13.2 \ dB$$

The directivity adjustment is:

$$\Delta_l = 1.9 \ dB$$

The corrected level is:

$$L_{ae}(P_1) = 97.4 + 7.0 - 13.2 + 1.9 = 93.1 \ dB$$

SEL at P(2), (1500,2000):

Sound exposure level at P(2) is determined by adjusting the reference level, corresponding to takeoff

power, for velocity and lateral attenuation.

$$L_{ae}(P_2) = L_{ae}(P, d) + \Delta_v - \lambda(0, d)$$

The aircraft speed is:

$$V = \sqrt{32^2 + (152.9^2 - 32^2)(1500/4408)} = 92.9 \ \text{nodi}$$

The speed adjustment for duration is:

$$\Delta_v = 10\lg(152.9/92.9) = 2.2 \ dB$$

Lateral attenuation is computed by:

$$\lambda(0,d) = 15.09\left(1 - e^{-1.67}\right) = 12.3 \ dB$$

The correct noise level is:

$$L_{ae}(P_2) = 99.1 + 2.2 - 12.3 = 89.0 \ dB$$

SEL at P3:

The point P(3) is beyond the point of lift-off and the ground track is straight, there are only two adjustment factors:

$$L_{ae}\ (P_3)\ =\ L_{ae}\ (P,\ d)\ +\ \Delta_v\ -\ \lambda\ (\beta,\ l)$$

Duration correction is:

$$\Delta_v\ =\ 10\lg\ (164/150)\ =\ 0.2\ dB$$

Lateral attenuation is calculated by:

$$\lambda\ (\beta, l) = (12.25)\ (5.28)\ /13.86 = 4.7\ dB$$

The corrected SEL is:

$$L_{ae}\ (P_3)\ =\ 99.1\ +\ 0.2\ -\ 4.7\ =\ 94.6\ dB$$

Segment	flight alt.	Path altitude	speed	thrust
takeof	0	0	0	EPR to
Initial climb	0-1000	4408-10454	152.9-152.9	8520
Flap retracts	1259	14658	180	8108
Continue climb	3000	28564	180	8430
Accelerate	3845	46907	250	7895

distance (meters)	takeoff thrust	climb thrust	cruise thrust
80	116.5	102.0	96.4
125	112.3	100.9	95.7
200	110.2	99.4	94.2
400	103.6	97.0	92.1
630	99.8	94.4	88.7

An appropriate aircraft noise assessment requires precise data on several levels, as it should account for the sources directly at the aircraft itself (level one) as well as flight paths (level two) and operation procedures (level three). This knowledge can be gathered for existing aircraft, but is hardly available during the design phase. Furthermore, aircraft design, operation procedures and other aspects might influence each other, as new aircraft designs might require new operation procedures or flight paths.

On the first level, the sound emission from aircraft components is computed based on flow computations around these components using computational fluid dynamics (CFD) methods. Typically, the results serve as input to codes which compute the aeroacoustics using computational aeroacoustic (CAA) methods. This procedure leads to high levels of prediction accuracy, but is restricted to small computation domains in comparison to a complete aircraft or even large scale scenarios with many aircraft on different flight paths. The next level is of greater scale but otherwise less detailed and aims for the prediction of the sound exposure on the ground for single aircraft. For this purpose, the sound emission of the dominant sound sources at the aircraft is being superimposed in order to form an overall aircraft radiation pattern. As the aircraft exhibits different flight configurations during a complete take-off-cruise-landing cycle, the radiation pattern is computed for all relevant configurations. On the last and coarsest level, radiation patterns are propagated to the ground for distinct flight paths and thus form a long-term noise footprint describing an entire scenario. Finally, noise footprints of such scenarios with many aircraft and other traffic types are superimposed in order to predict the overall noise exposure of residents on large time scales, for example a year of operation.

Within this contribution, a multi-level, multi-fidelity approach is presented that inherently couples domain-specific numerical tools of all three aforementioned levels of detail and thereby composes a chain, which enables the prediction of noise exposure on the ground due to present and future aircraft including sound propagation from the source to the receiver. With this method available, new low noise aircraft design technologies become assessable at the receiver location using perception-focused quantities. This possibility provides a great benefit for aircraft design regarding the goals of "Flightpath 2050".

On the first level, this chain is composed of the CFD code TAU, developed at DLR (DLR—German Aerospace Center (Deutsches Zentrum für Luft- und Raumfahrt)), and CAA code PIANO (Perturbation Investigation of Aerodynamic NOise), developed at DLR, for computing the noise emission on component level. The tool PANAM (Parametric Aircraft Noise Analysis Module), developed at DLR, computes the noise emission of entire aircraft for different flight configurations in level two. In level three, the aircraft noise simulation tool sonAIR, developed at Empa, is used to perform the calculation of sound propagation as well as to superimpose the noise of many single aircraft in entire scenarios.

Noise Assessment of an Aircraft Component (Level 1)

Noise assessment of aircraft components can be accomplished using numerical and experimental investigations. Typically, for new technologies no measured aeroacoustic data is available on aircraft components. Furthermore, the effort to set up e.g., comprehensive aeroacoustic wind tunnel tests usually is very high, slow and costly. Semi-empirical source models, as used for system noise prediction tools, are typically based on some theoretical concept and a calibration with measured data. For the lack of such data and in view of a seamless numerical assessment of new technologies, it is mandatory to provide the relevant aeroacoustic database for a technology on the basis of efficient, high fidelity CAA simulations.

For the work presented, a CAA concept is pursued, which is non-empiric because it does not assume any preset geometry or component, thus general enough on the one hand and very efficient due to modeling on the other hand, to provide fast enough response times. This concept is hybrid in the sense that initially a RANS (Reynolds-averaged Navier-Stokes) flow field simulation is done with DLR's CFD Code TAU. The second step consists of two parts, (a) the stochastic unsteady 3D modeling of the turbulence by DLR's stochastic turbulence generator fRPM (fast Random Particle Mesh Method) and (b) the actual CAA computation of the (acoustic) perturbation field about the given RANS flow field by means of DLR's CAA code PIANO.

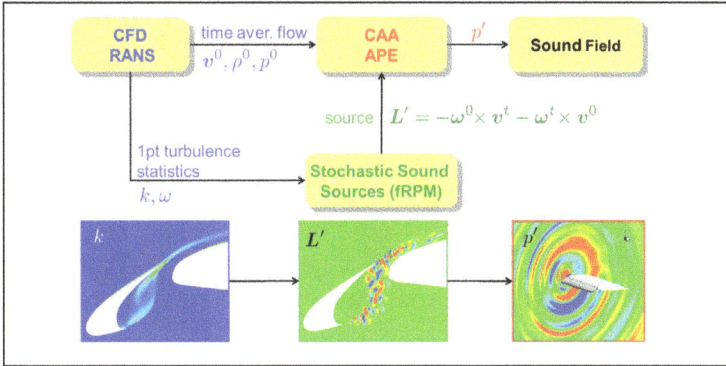

CAA component noise prediction concept, on the example of slat noise.

The approach requires an appropriate modeling of the unsteadiness of turbulence, which means that it has to cover those features of the turbulence which are relevant for the sound generation. The respective turbulence representation is realized with the stochastic turbulence generator "fRPM" which takes the flow field and one point turbulence statistics from a pre-cursor RANS mean flow simulation as an input. The convection characteristics are met correctly, while the given distribution of the turbulence kinetic energy k, and a local correlation length is satisfied.

Turbulence is only realized in the relevant source domain (called "fRPM source patch"). The source region is discretized by a Cartesian mesh with an appropriate resolution of Dx to resolve the turbulent structures which determine the number of mesh points, typically ranging around numbers of 10^5 for 2D simulations. As discussed in the dynamic of the synthesized turbulence is represented by an ensemble of particles, moving along in the (given) mean flow provided by the RANS CFD solution. The number of particles required to represent the turbulent eddies typically is slightly larger than the number of mesh points in 2D. The mesh for the acoustic propagation is set to resolve acoustic wavelengths of the desired maximum frequencies with 7 points per wavelength.

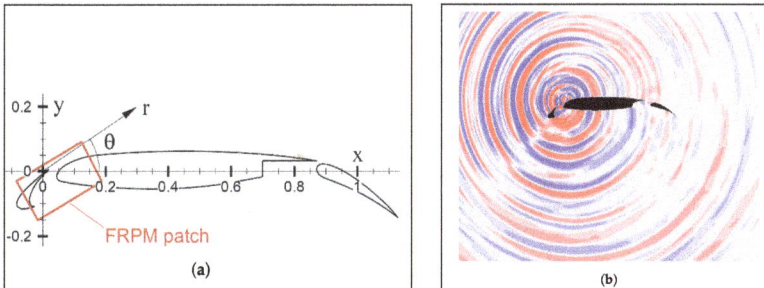

Typical computation setup for CAA simulation of slat noise and results of the pressure distribution.
(a) Arrangement of fRPM (fast Random Particle Mesh Method) source patch (red box);
(b) Snapshot of acoustic pressure (red=positive fluctuation, blue=negative fluctuation).

Although the turbulence is stochastically realized in 3D, for 2D geometries like a slat the CAA computations may be carried out in a wing section in 2D since spanwise velocity fluctuations do not contribute to the generation of slat noise. This approach enables very fast computation and thus short turnaround times (a few hours on a work station) to determine spectral differences of systematically varied slat configurations. In general one may state that this approach is valid for any flow whose time average can be described appropriately based on the RANS flow model.

Level 1 Interface to Level 2 (Full Aircraft Prediction)

The predicted sound pressure spectra of design variants are related to the spectrum of the respective (standard) reference aircraft component. The resulting spectral difference is transferred to the level two predictions. This "delta" information is used to accordingly modify the level two prediction model of that component as a function of observer orientation and flight condition, implemented as a look-up table. In case that a low noise technology affects the aerodynamics of the aircraft, respective corrections are to be extracted from the level one simulations and transferred to level two as well.

Validity of Level 1 Prediction for Slat Noise Reduction Technology

Noise prediction concept is demonstrated on the example of the noise generated at the leading edge of a transport aircraft's wing in high lift configuration, i.e., when the leading edge slats and the trailing edge flaps are deployed for landing. During this phase of flight, the high lift system of a modern, say Airbus A320 type, transport aircraft is a very important source of community noise. In fact, together with the deployed landing gear, the slat can dominate the aircraft overall noise at the certification location "approach". This is particularly true for aircraft equipped with high bypass ratio turbofan engines (e.g., A320 NEO).

Slat noise is a very difficult problem to simulate numerically, since its origin is the presence of turbulence in the strongly accelerated flow through the slot between slat and main element. Low noise targeted design modifications are practically impossible to carry out with scale resolving approaches such as "Large Eddy Simulation" (LES) since the number of required simulations would lead to unrealistic computation times.

In the following, the validity of the CAA approach for slat noise is demonstrated by comparison to acoustic data, measured in DLR's AcousticWind tunnel Braunschweig (AWB). The results of a systematic slat gap and overlap variation (definitions) are shown. The low noise slat setting was found by simulation of a large set of gap and overlap positions. The reference and the variant exhibiting least noise were implemented into DLR's research high lift airfoil model F16 with a chord length of 300mm and a span of 800mm in AWB. The prediction of the one-third octave band spectra corresponds well with the measured result and the difference in spectra between reference and the low noise setting are predicted very accurately. Only in the very low frequency domain the results differ due to parasitic noise originating from the wind tunnel background noise. The data shows that the computation concept for slat noise is valid and may be used for different slat designs. In topic this method will be used to predict the noise reduction achieved by re-designing the slat to a so called "very long chord slat" (VLCS).

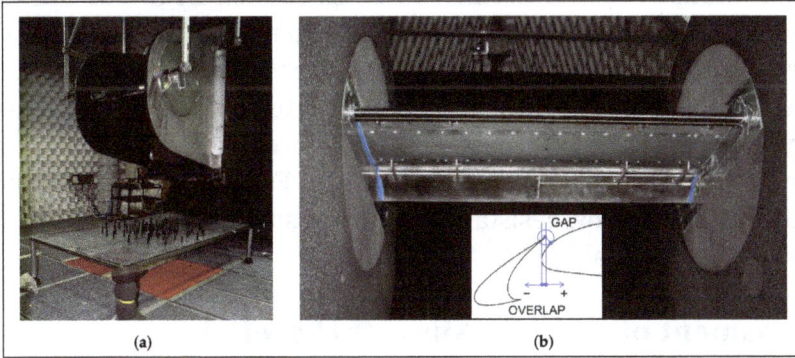

AWB (AcousticWind Tunnel Braunschweig) test arrangement for high lift airfoil F16 and detailed view of installed F16 (slat in front). (a) F16 airfoil in AWB (side view); (b) F16 airfoil pressure side in AWB (view from upstream); defintion of overlap and gap.

A second noise reduction technology for high lift noise, which will be implemented in the overall assessment is the concept of a so called "slat cove liner". In this case the noise reduction is not accomplished by manipulating the local flow and therefore the sound generation. Instead, present slat sound is absorbed by a liner, placed in the near-field of the slat source, namely in the cove area. For this reason, this technology will be further referred to as "slat cove liner".

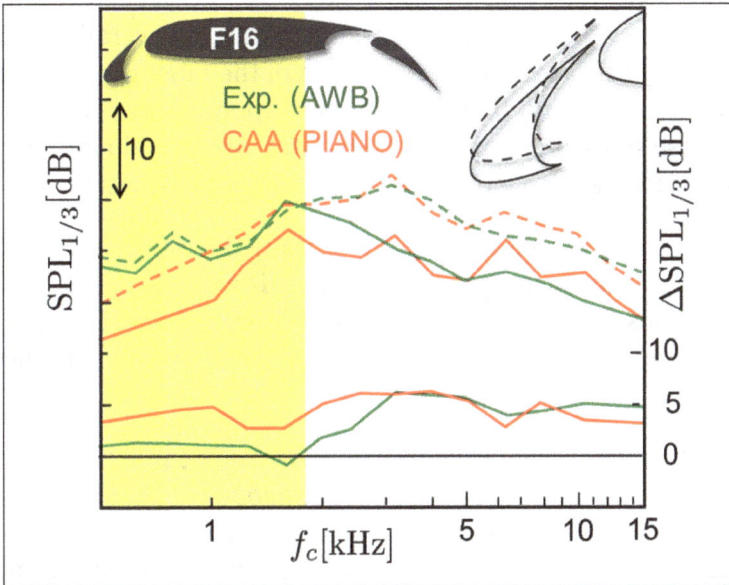

Slat noise prediction and comparison with AWB experimental data including slat setting variation; dashed lines correspond to reference slat setting, solid lines refer to low noise slat setting. Right hand scale referring to spectral difference. Yellow shade area contaminated by wind tunnel background noise.

Although this noise reduction technology was found experimentally on the F16 slat, its application to different high lift leading edge devices is of high interest. In other words, not only the source noise reduction by the shaping of the slat is to be covered by the numerical prediction approach, but simultaneously the effect of a cove liner. A respective prediction capability was developed within the German Corporate Research Center CRC880, funded by the German research council DFG. For this purpose, the equations solved within the prediction chain are extended by extra terms which model the presence of porous volumes. DLR's RANS solver TAU and perturbation solver PIANO were re-formulated in volume-averaged form to describe the presence of porous material in a homogenized way. This results in the occurrence of extra modeling terms in the momentum equations, so called DARCY and FORCHHEIMER terms, involving the permeability (inverse flow resistance) and porosity (volume fraction of air of connected pores) of the material.

Noise Assessment of a Single Aircraft (Level 2)

In the context of the evaluation of acoustic properties of novel aircraft technology, the term aircraft system noise is used. According to BURLEY aircraft system noise is defined as the "source noise emitted by an aircraft in flight, propagated through the atmosphere, and received by observers or sensors on the ground". Consequently, the aircraft noise emission has to be simulated in a first step. As the noise emission of separated sources is covered by the numerical investigations in level one as, level two provides the overall noise radiation patterns of single aircraft for every necessary flight configuration by combining the single sources of level one. Within this contribution, this is done using the DLR-tool PANAM.

When simulating the noise emission, it is assumed that the overall aircraft noise emission can adequately be approximated as the sum of the most relevant individual noise sources/components on-board, thus a component approach is pursued. Thereby, individual relevant noise sources are modeled separately and certain interaction effects are accounted for to finally assemble the overall aircraft noise emission.

The selected noise source models are parametric with respect to the aircraft/engine design parameters and the aircraft operation. This is a prerequisite if different vehicle concepts along various flight procedures are to be investigated. The quantity and complexity of the required input data is still manageable and available in the early conceptual design phase.

The overall aircraft noise is separated into airframe and engine noise contribution. Selected airframe noise source models have been developed by DLR as described in previous literature. These DLR in-house models describe the major noise sources and some dominating interaction effects. Explicit models for the simulation of major turbofan engine noise sources are available from literature. PANAM predicts turbofan engine noise as a combination of the two dominating noise sources jet and fan, respectively.

Stone's model is employed for the jet noise originating from the core and bypass jet. For the fan noise simulation the Heidmann model is applied. The empirical database of the original Heidmann model has been modified in order to reflect more recent developments in the field of engine fan design. Possible shielding effects of the engine fan noise are considered using the DLR ray-tracing tool SHADOW. The application of SHADOW allows studying promising low noise aircraft concepts with significantly reduced fan noise contribution.

PANAM usually is applied to perform the aircraft system noise prediction for all selected aircraft design concepts along their individual flight trajectories. Within the presented simulation process, PANAM is applied to predict emission levels of single aircraft as input for level three. Furthermore, the ground noise predictions of PANAM are used to optimize the flight trajectory of each aircraft which is also a required input to level three. Furthermore, PANAM can be included as a module within the PrADO (Preliminary Aircraft Design and Optimisation Program) simulation process so that all the required input data for a noise prediction can be generated as described by Bertsch. At this point, the process can readily be applied towards medium-range, conventional transport aircraft with turbofan or propeller engines.

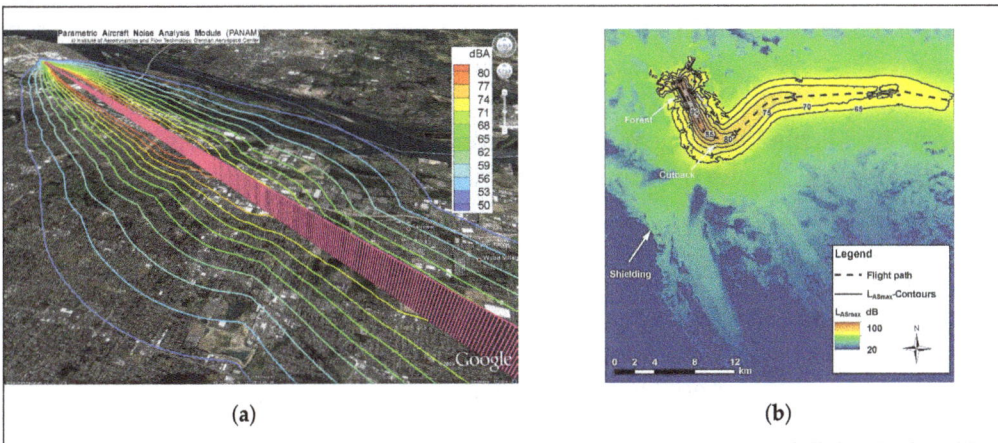

(a)   (b)

In figure (a) Noise exposure footprint of an approaching aircraft, computed using PAN-AM (Parametric Aircraft Noise Analysis Module). The isolines refer to the A-weighted ground noise level; (b) Acoustic footprint (LA,S,max) of an A330-300 take-off on runway 16 of Zurich airport. Specific source and propagation effects are labeled.

Noise Assessment of System Scenarios (Level 3)

In this step, the authors study the effect of new technologies within entire airport scenarios with the scope to minimize the noise distribution on ground. For this purpose, the aircraft noise simulation tool sonAIR is applied. The tool comprises of a sound source database and a sound propagation model, which are formulated for one-third octave bands within a frequency range of 25 Hz to 5 kHz.

The sound source database was developed for fast calculations using the sound emission model, which predicts the sound emission of turbofan powered aircraft as a function of the flight configuration (power setting, Mach number and aeroplane configuration, i.e., flaps, slats and landing gear). The database consists of two look-up tables for airframe and engine noise sources, where the sound emission levels are stored as a function of frequency, directivity and flight parameters. This interface is used to implement PANAM outputs from the second level to sonAIR. Thus, the description of the sound source is still very detailed and the loss of information is small.

For the calculation of the sound propagation, a hybrid modeling approach is implemented in sonAIR. In case of simple situations, where direct sound is dominant, only geometric divergence and atmospheric absorption and a mean ground effect are accounted for. In other cases, where the sound incidence angles are low, the sound propagation model sonX can be applied to account for effects as shielding (due to e.g., buildings, topography), ground reflections or foliage attenuation. However, for the proof of concept the sound propagation is simplified to direct sound as the topography is simplified, too.

For the task of noise mapping, calculations are performed for a receiver grid with a constant height above the ground. In aircraft noise simulations, typically numerous single flights are processed together resulting in a footprint, i.e., the sound exposure averaged over a bundle of flights, on all receiver points arranged in a grid. An example of a noise footprint during take-off is given, indicated by the LA,S,max. However, the quantity to be considered within this contribution is the LA, E, meaning the A-weighted exposure level. As a last step, these footprints are weighted with the number of movements per aircraft and route and the contributions from all aircraft and route combinations are energetically summed up.

The sonAIR-software is embedded into a geographic information system (Esri ArcGIS), in which the input data as well as the results are stored and visualized. Such environment also facilitates comparisons of noise contours or statistics on the number of affected people (e.g., high annoyance, sleep disturbance).

References

- Heavyside's Approach to an Elliptical PDE in a Significant Physical Problem, 5th WSEAS / IASME International Conference on Engineering Education (EE'08), Heraklion, Crete Island, Greece, July 22-24, 2008. ISBN: 978-960-6766-86-2 / ISSN 1790-2769

- Transport-noise: umweltbundesamt.de, Retrieved 15 March, 2019

- Infinitesimal Equivalence between Linear and Curved Sources in Newtonian Fields: Application to Acoustics, International Journal of Mechanics, Issue 4, Vol.1, pp. 89-91 (2007), ISSN: 1998-4448

- An Operatorial Approach to Sturm-Liouville Theory with Application to the Problem of a Spherical Conductor Embedded in a Uniform Field, International Journal Of Mathematical Models And Methods In Applied Sciences, Issue 2, Vol.2, pp 285-293 (2008), ISSN 1998-0140

- Aircraft-Noise-Assessment-From-Single-Components-to-Large-Scenarios-323144284: researchgate.net, Retrieved 14 April, 2019

Chapter 3

Industrial and Construction Noise

Industrial machinery and processes are composed of various noise sources such as rotors, stators, gears, fans, vibrating panels, turbulent fluid flow, electrical machines, internal combustion engines, etc. This chapter closely examines the different sources of industrial and construction noise to provide an extensive understanding of the subject.

Industrial Noise

Industrial machinery and processes are composed of various noise sources such as rotors, stators, gears, fans, vibrating panels, turbulent fluid flow, impact processes, electrical machines, internal combustion engines etc. The mechanisms of noise generation depend on the particularly noisy operations and equipment including crushing, riveting, blasting (quarries and mines), shake-out (foundries), punch presses, drop forges, drilling, lathes, pneumatic equipment (e.g. jack hammers, chipping hammers, etc.), tumbling barrels, plasma jets, cutting torches, sandblasting, electric furnaces, boiler making, machine tools for forming, dividing and metal cutting, such as punching, pressing and shearing, lathes, milling machines and grinders, as well as textile machines, beverage filling machines and print machines, pumps and compressors, drive units, hand-guided machines, self-propelled working machines, in-plant conveying systems and transport vehicles. On top of this there are the information technology devices which are being encountered more and more in all areas.

Industrial Noise Sources

The fundamental mechanisms of noise sources are discussed, as well as some examples of the most common machines used in the work environment. The sound pressure level generated depends on the type of the noise source, distance from the source to the receiver and the nature of the working environment. For a given machine, the sound pressure levels depend on the part of the total mechanical or electrical energy that is transformed into acoustical energy.

Sound fields in the workplace are usually complex, due to the participation of many sources: propagation through air (air-borne noise), propagation through solids (structure-borne noise), diffraction at the machinery boundaries, reflection from the floor, wall, ceiling and machinery surface, absorption on the surfaces, etc. Therefore any noise control measure should be carried out after a source ranking study, using identification and quantification techniques. The basic mechanism of noise generation can be due to mechanical noise, fluid noise and electromagnetic noise.

The driving force for economic development is mainly the endeavour to produce consumer goods ever more cost-effectively. From the point of view of the machine manufacturer, this generally means offering products with a low space, material, energy and production time requirement (smaller, lighter, more economical and more productive). At the same time account is being taken increasingly of resource conservation and environmental friendliness, although the rise in noise levels which frequently goes along with increased output and productivity is often overlooked. Personnel are then exposed to higher noise levels than before, despite noise-reducing measures taken in the machine's design. This is because the noise emission rises non-linearly because of higher rotary and travelling speeds in machine parts.

For example, for every doubling of the rotary speed the noise emission for rotating print machines rises by about 7 dB, for warp knitting looms 12 dB, for diesel engines 9 dB, for petrol engines 15 dB and for fans is between 18 to 24 dB. For the purpose of comparison: the doubling of sound power produces an increase in emission of 3 dB only.

But even previously quiet procedures are often replaced by loud ones for reasons of cost, e.g. stress-free vibration instead of annealing for welded parts. In some cases new technologies also result in higher emissions; for example, with the use of phase-sequence-controlled electrical drives, the excitation spectrum shifts further to high frequencies, which results in a greater sound radiation from large machine surfaces. This means that some new noise problems are closely related to the use of modern technologies.

Mechanical Noise

A solid vibrating surface, driven or in contact with a prime mover or linkage, radiates sound power (W in Watts) proportional to the vibrating area S and the mean square vibrating velocity < v² >, given by;

$$W = \rho c S \left\langle v^2 \right\rangle \sigma_{rad}$$

Where,

ρ is the air density (kg/m³),

c is the speed of sound (m/s) and

σ_{rad} is the radiation efficiency.

Therefore care must be taken to reduce the vibrating area and reduce the vibration velocity. Reducing the vibrating area can be carried out by separating a large area into small areas, using a flexible joint. Reduction of the vibration velocity can be carried out by using damping materials at resonance frequencies and blocking the induced forced vibration. A reduction of the excitation forces and consequently of the vibration velocity response by a factor of two can provide a possible sound power reduction of up to 6 dB assuming that the other parameters are kept constant. Typical examples of solid vibration sources are: eccentric loaded rotating machines, panel and machine cover vibration which can radiate sound like a loudspeaker, and impact induced resonant free vibration of a surface.

Fluid Noise

Air turbulence and vortices generate noise, especially at high air flow velocities. Turbulence can be generated by a moving or rotating solid object, such as the blade tip of a ventilator fan, by changing high pressure discharge fluid to low (or atmospheric) pressure, such as a cleaning air jet or by introducing an obstacle into a high speed fluid flow.

The aerodynamic sound power generated by turbulent flow is proportional to the 6th to 8th power of the flow velocity ($W \sim U^{6 \text{ to } 8}$), which means that a doubling of the flow velocity (U) increases the sound power (W) by a factor of 64 to 254 or 18 to 24 dB respectively. Table shows the effects of doubling of the typical velocity together with other primary mechanisms. Therefore care must be taken to reduce flow velocity, reduce turbulence flow by using diffusers and either remove obstacles or streamline them.

Examples of Machinery Noise Sources

Here, noise sources are presented for the most common machines used in industrial installations. For each case, the mechanism of noise generation is discussed.

Industrial Gas Jets

Industrial jet noise probably ranks third as a major cause of hearing damage after that of impact and material handling noise. Air jets are used extensively for cleaning, for drying and ejecting parts, for power tools, for blowing off compressed air, for steam valves, pneumatic discharge vents, gas and oil burners, etc. Typical sound pressure levels at 1 m from a blow-off nozzle can reach 105 dB(A).

Table: Increase of noise given by the sound power level difference ΔL_w due to doubling of typical velocity (e.g. average flow velocity of gas jets, rotational speed of fans).

Mechanism	Example	Increase in sound power due to doubling typical velocity
Pulsation	Reciprocating compressor,	12 db
Turbulence	Exhaust fan	18 db
Jet	Compressed air expansion	24 db

Reservoir compressed air pressure is usually in the range of 45 to 105 psi (300 to 700kPa). The air acceleration varies from near zero velocity in the reservoir to peak velocity at the exit of the nozzle. The flow velocity through the nozzle can become sonic, i.e. reaches the speed of sound. This results in a high generation of broad-band noise with the highest values at a frequency band between 2 to 4 kHz.

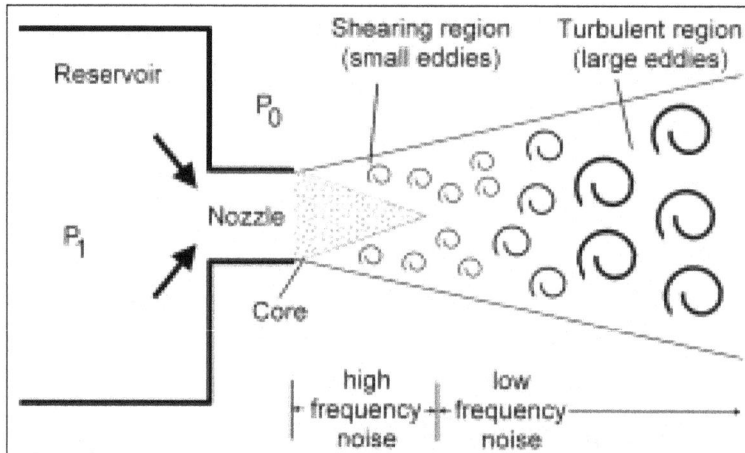

Noise sources in gas jet.

The mechanisms of generation of the noise from gas jets results from the creation of fluctuating pressures due to turbulence and shearing stresses as the high velocity gas interacts with the surrounding medium. High and low frequency bands of noise are formed, due to the complex radiation sources; high frequency noise is generated near the exit nozzle in the mixing region and the low frequency noise is generated downstream at the large scale turbulence. Therefore, the spectral character of gas-jet noise is generally broadband.

Ventilator and Exhaust Fans

It is rare not to find one or more ventilators or exhaust fans in each department of an industrial or manufacturing complex. Fan and blower noise is the easiest and most straightforward noise problem to solve, using an absorptive type silencer.

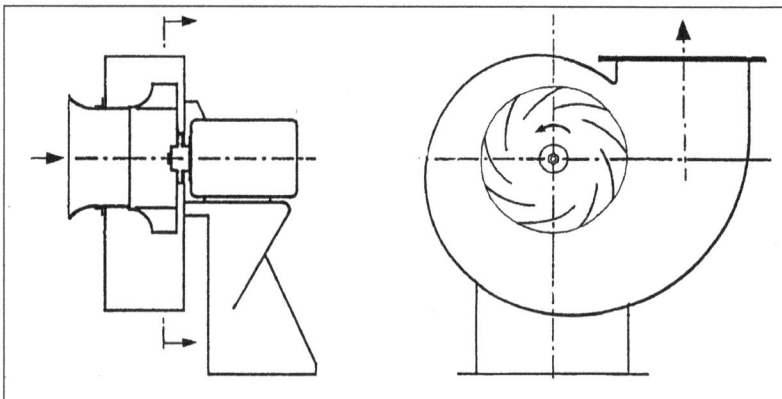

Example of a centrifugal fan, rotor with backward-curved blades.

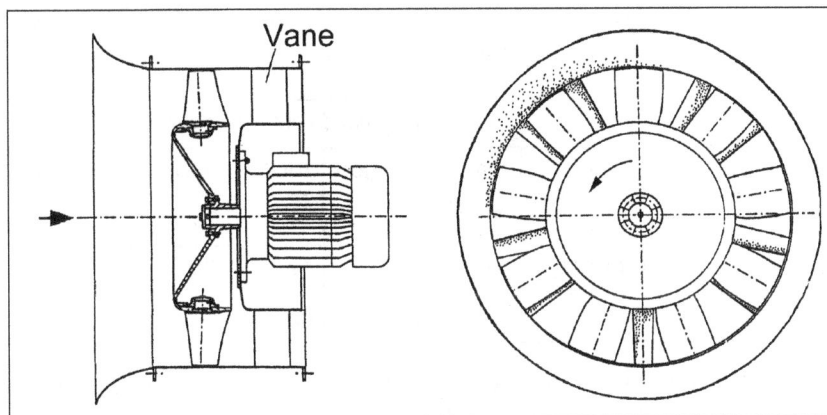

Example of a vaneaxial fan.

Fans are used to move a large volume of air for ventilation, by bringing in fresh air from the outside, blowing out dust, vapour or oil mist from an industrial environment, and for a drying or cooling operation, etc. Industrial fans are usually low-speed, low-static-pressure and have a large volume flow rate. Ideally, fans should operate at the maximum efficiency point on the pressure flow curve characteristic. Therefore, the choice between axial or centrifugal fans is made by the manufacturer to satisfy maximum efficiency at a certain static pressure/flow rate.

Three basic noise sources are:

- Broadband aerodynamic noise generated by the turbulent flow.

- Discrete tones at the blade passing frequency F_p (Hz) given by: F_p = (Rotation in RPM x Number of blades/60), and the harmonics ($2F_p$, $3F_p$, etc).

- Mechanical noise due to mounting, bearing, balancing, etc.

The sound power level (L_w) generated by fans (without the drive motor) can be easily predicted in the early project stages of an industrial installation using the Graham equation for each of the octave bands from 63 to 8000 Hz.

$$L_w = K + 10\log_{10}\overline{Q} + 20\log_{10}P_a + C \quad dB$$

Where Q is the flow rate (m³ /sec), P_a is the static pressure (kPa), K is the specific sound power level for each of the octave bands based on a volume flow rate of 1 m³/s and a total pressure of 1 kPa and C is a constant to be added only at the octave band containing the blade passing frequency.

Based on the sound power predicted by the above equation, the sound pressure levels can estimate at specified locations in certain installations. The finite element, boundary element or ray acoustics methods are available in commercial software programs for these estimates or a simplified diffuse field model can be used for sound pressure level estimate.

Table: Specific octave band sound power levels K in dB(re 1 pW) of three types of fans with wheel size under 0.75 m based on a volume flow rate of 1m³/s and a total pressure of 1 kPa.

Fan type	Octave band center frequency [Hz]								C
	63	125	250	500	1000	2000	4000	8000	
Radial, backward-curved	90	90	88	84	79	73	69	64	3
Radial, straight blades (no figure)	113	108	96	93	91	86	82	79	8
Vaneaxial, hub ratio 0.6-0.8	98	97	96	96	94	92	88	85	6

The table gives average values which widely scatter due to the properties of the complete system with ducts. The column "C" contains minimum values which even in the case of the least noisy fan with backward-curved blades may be sometimes double as high.

Compressors

Compressors are usually very noisy machines with high pressure. There are several types of compressor: rotary positive displacement (lobed impellers on dual shafts), gear or screw compressors, reciprocating compressors and liquid ring compressors are the most common.

The basic noise sources are caused by trapping a definite volume of fluid and carrying it around the case to the outlet with higher pressure. The pressure pulses from compressors are quite severe, and equivalent sound pressure levels can exceed 105dB(A). The noise generated from compressors is periodic with discrete tones and harmonics present in the noise spectrum.

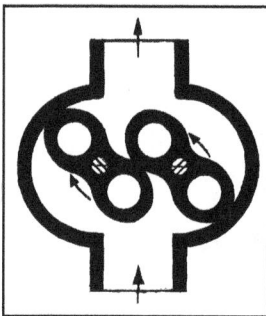

Rotary Positive Displacement Compressor.

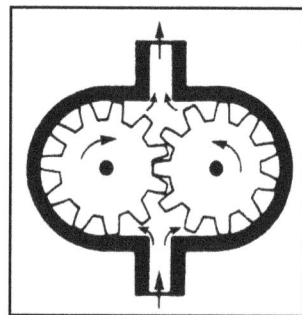

Gear Compressor.

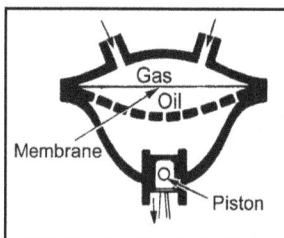

Reciprocating Compressor.

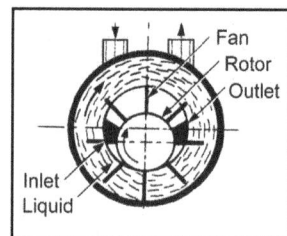

Liquid Ring Compressor.

Electric Motors

Noise from electrical equipment such as motors and generators is generally a discrete low frequency, superimposed on a broadband cooling system noise. The electric motor converts electrical energy to magnetic and then mechanical energy with the output of a useful torque at the motor shaft. Part of the energy transformation is converted to heat, causing a rise in rotor, stator and casing temperature; therefore an electric motor must be supplied with a cooling fan system. The cooling fan can be incorporated inside as in the case of an "OPEN" motor or outside as in the case of a "Totally Enclosed Fan Cooled (TEFC)" motor. TEFC motors are more widely used, due to their robust construction which can withstand a dirty environment. OPEN motors are less used due to possible contamination by the environment. An OPEN motor is sometimes (but not always) less noisy than a TEFC motor since the noisy fans are incorporated inside.

There are three basic sources involved in the noise generated by electric motors:

- Broad-band aerodynamic noise generated from the end flow at the inlet/outlet of the cooling fan. The cooling fan is usually the dominant noise source.

- Discrete frequency components caused by the blade passing frequencies of the fan.

- Mechanical noise caused by bearing, casing vibration, motor balancing shaft misalignment, and motor mounting. Thus careful attention should be given to the vibration isolation, mounting and maintenance.

Noise generated by the motor fan is the dominant motor noise source, especially for TEFC motors. A sharp increase in noise occurs as the shaft rotational speed increases from 1800 to 3600 RPM. For large motors in the range of 1000 kW, 3600 RPM, a sound pressure level of as high as 106 dB(A) occurs. Measurements carried out in the laboratory for a range of TEFC motors from 25 to 2500 HP, no load, with and without the straight blade motor fan, show a difference of up to 50 dB(A) in the total sound pressure level. This large distribution of the fan noise is due to the fan shape. Motor fan blades are usually straight, so that the motor cooling is independent of rotation direction. Straight blade fans are very noisy, due to the large aerodynamic turbulent sound generated. Noise reduction in electric motors can be achieved by the use of an absorptive silencer or by redesign of the cooling fan, e.g. with irregular spacing of straight blades.

Woodworking Machines

The woodworking industry has experienced noise level increases as a result of modern, higher speed, and more compact machines. The basic noise elements in woodworking machines are cutter heads and circular saws. Equivalent sound pressure levels (L_{Aeq}) in the furniture manufacturing industry can reach 106 dB(A).

Woodworking machinery uses operations, such as cutting, milling, shaping, etc. Three basic noise sources are involved:

- Structure vibration and noise radiation of the work piece or cutting tool (such as a circular saw blade) and machine frame, especially at the mechanical resonance frequencies.

- Aerodynamic noise caused by turbulence, generated by tool rotation and the workplace in the air flow field.

- Fan dust and chip removal air carrying systems.

Pneumatic Tools

Compressed air-powered, hand-held tools such as drills, grinders, rivetting guns, chipping hammers, impact guns, pavement breakers, etc. are widely used within a broad spectrum of different industries. There are three basic types of sources that dominate the noise generated:

- Noise produced by contact between the machine and the working surface. The vibration transmitted from the tool tends to vibrate the working surface and work bench, generating high radiation noise, especially at mid and high frequencies.

- Exhaust air noise caused by the turbulent flow generated as the compressed air passes the motor and by the aerodynamic noise generated in the air exhaust.

- Sound radiation from tool vibration caused by air flow inside the tool.

The noise level of hand held tools can reach as high as 110 dB(A) at the operator's ear.

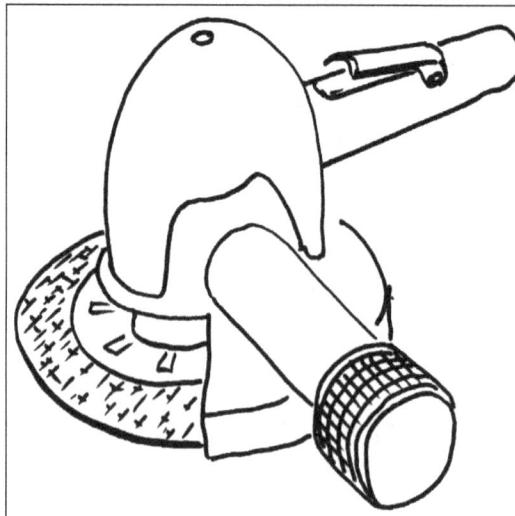

Typical grinding air powered hand tool.

Typical Noise Levels

As an example, data collected in Singapore in 1993 shows that only 366 factories out of 9051 factories have a hearing conservation program implemented. Table shows a list of factories and number of workers with and without hearing conservation program implemented. Table shows a list of 20 factories with a range of sound pressure levels and an average level for each. Data collected in Denmark on 55 pneumatic and electric hammers in different industries show an SPL of between 88 to 103 dB(A).

Planing wood machine operators are exposed to an SPL maximum of 101 dB(A) and an SPL minimum of 96 dB(A) with an L_{Aeq} = 98 dB(A) for 8 hours, which is far above the acceptable risk values (AIHA-USA).

Data collected in a cigarette factory in Brazil show an SPL level of a compressed air cleaning process of up to 103 dB(A), with an L_{Aeq} = 92 dB(A) for 8 hours.

Economic calculations have shown that administrative and technical preventive measures are profitable. Technical progress during recent years has led to a decrease of the very high noise exposures, but not much change in moderate and low noise exposures. Measurements taken at some typical occupations show that the L_{Aeq} levels in the occupations shown by experience to have the worst noise have been 88-97 dB with highest peak levels of 101-136 dB. In the referenced study, there were no findings of peak levels exceeding 140 dB.

Noise Maps for Industrial Sources

Noise evaluations around industrial premises have been for a long time carried out by means of short term noise measurements, on a limited number of receivers, and by performing some simplified qualitative analysis, eventually with some simple calculations. However, experience has demonstrated that often, and above all in installations of large dimension and complexity, this approach does not enable one to get feasible results neither a clear vision of the real noise impact. Moreover, it does not, in general, produce enough information for decision making with respect to what noise control measures should be implemented – has it does not enable one to identify and rank the noise sources – and it does not enable one to predict the noise impact of a new factory or of this or that noise control action on an existing installation. Other non-negligible aspect consists on the difficulty that "traditional" noise assessments have in presenting results which are easily apprehended by non-specialists. This makes it difficult an effective communication of the results attained by noise control programmes to the potentially interested publics, such has the neighbouring communities, governmental agencies, environmental inspectors and auditors, shareholders, insurance companies, municipalities, environmental NGOs. Thus, all the effort and investment put by the organization on reducing its noise emissions is often not recognized by those interested parts, losing the opportunity to enhance the image of the company.

The development of computer modelling techniques that simulate the acoustic emission and propagation enables one, in our days, to model with good accuracy and reasonably fast, the most complex scenarios of noise generation and propagation. The results are normally presented in the form of coloured noise maps, each colour corresponding to a given interval of noise levels, typically in steps of 5 dB. Above all, such a model, if correctly developed, enables one to get a true noise monitoring and management system, from which it is possible to rank noise sources, extract the individual contributions of each noise source to any given receiver, update the information whenever changes are introduced in the factory, and establish detailed noise control action plans and predicting its results. Difference maps can also be easily extracted from such a model, in order to depict before vs. after maps, total noise vs. background noise maps, or scenario 1 – scenario 2 maps, etc.

The need for an industrial noise map can come up to on a wide range of situations: environmental impact assessment for the installation of a new factory or for changing an existing one, environmental license such as under IPPC regulations, and complaints from neighbours, certification under ISO 14000 or EMAS, where a full demonstration of compliance with noise regulations is required. In certain cases, such as industrial parks, the aim can be to control noise build up during the successive installation of new industries in the park, which can be done by establishing noise quota for each lot in the park.

Whichever the reason, the number one aim on producing a noise map – and above all the acoustic model on which it is based – should be to get a useful tool which will enable one to correctly evaluate a noisy situation, be it already existent or planned for the future, and study what the best solutions are to comply with given noise limits around the plant. By ranking the sources and predicting the practical outcome of any scenario, one can effectively optimize the investment in noise control actions.

General Methodology for Noise Mapping of Industrial Plants

The general methodology for noise mapping of industrial plants can be resumed as illustrated.

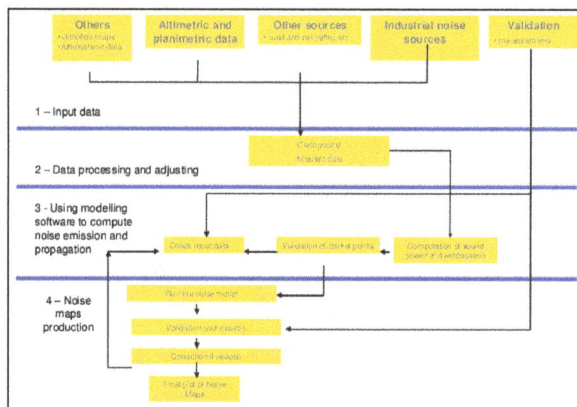

General methodology for noise mapping of industrial plants.

Obtaining correct input data is, as in any model, the most critical part of the job: the "garbage in, garbage out" expression fully applies here.

Moreover, in industrial noise mapping projects, this is also the hardest part of the job, as very seldom one has access to accurate digital 3D drawings of the plant or, even less, to adequate acoustical data such as sound power level or directivity of the noise sources. Therefore, in most cases of existing plants, one must get all these crucial data the "hard way", which means:

- Actually spending several weeks on site in order to understand, as deeply as possible, how the industry actually works;

- Getting close to all sound sources in order to measure its noise appropriately;

- Draw by scratch all industrial buildings to insert them into de model;

- Decide how to model each noise source;

- Accurately position them in the drawings in order to be able to insert them at their right places in the model.

It is worth noting that the sources can easily add up to more than one hundred and, sometimes, can come close to a thousand. Also, for interior sources, one has often to estimate, or actually measure, the transmission loss of the building elements involved in the in-out propagation process.

Getting input data for the acoustic model of a factory: Close field measurements.

After collecting all required data, this is introduced in the model and this usually means many adjustments, both for the geometrical data – by checking all information, taking advantage of 3D visualization capabilities of modern noise modelling software, such as CadnaA v3.7 – and for the acoustical data – generally by running calculations at a number of control points corresponding to real points where validation measurements have been taken, and comparing the measured versus calculated values. This is an iterative process which normally means going back into the field to check out doubts, make new measurements and have meetings with engineers from the factory to verify that this or that machine has been running on its normal condition, etc.

Source noise data is normally introduced in the model in the form of octave band sound power level, accounting for directivity and for % working time for each reference period of the day, if relevant for the project. When data comes from actual measurements on site, either sound power levels have been actually taken (e.g. using sound intensity measurements) or, as is normally the case, an estimation of the sound power level is made from pressure measurements close to each source, taking into account its mounting conditions, the presence of reflecting surfaces or other machines nearby and performing a validation process by comparing measured versus calculated noise levels at positions further apart from the source. In fact, full sound power determination for each source according to standards such as the ISO 3740 series, or ISO 9614 is generally out of the question, except in some special situations, such as checking a new machine during set up process or start-up of a new factory, and the like.

Simplified methods are therefore used, and care must be taken to ensure one makes the right simplifications and that the validation procedure is extensive enough so that your successive iterations can make the model converge into an accurate acoustic model. This validation process normally encompasses two steps:

- Source validation: Where single sources or small groups of sources are validated by means of checking measured against calculated values on a set of receivers at intermediate distances, not too close but not too far away from the sources which are being validated, normally inside the plant perimeter, but in the acoustic far field of each individual source.

- Full model validation: Where the entire plant, with all its sources, is validated by means of checking measured against calculated values on a set of receivers far away from the sources, typically outside the plant perimeter and sometimes close to sensitive receivers, such as neighboring dwellings.

When it is possible (or it just happens) to stop some machines or groups of machines, one can take advantage of that to facilitate the process of getting and validating noise data.

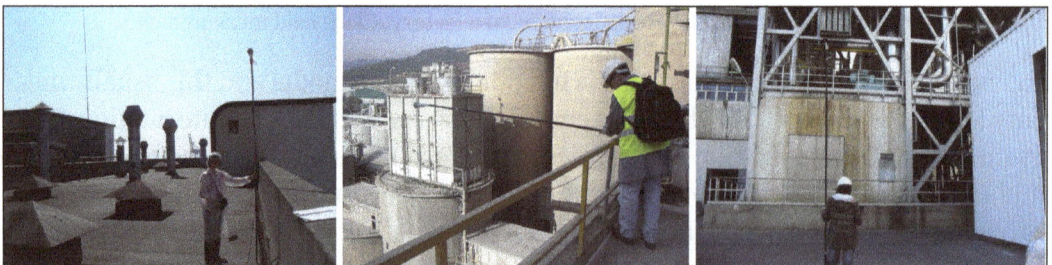

Source validation: Measurements within the factory but not too close to the sources.

Model validation: Measurements outside the factory, often close to sensitive receivers.

Ofcourse, when dealing with a project for a new factory, things work out differently, and one has to try hard to find acoustical data from suppliers of all types of machines or, when unavailable (as happens too much often), rely on available literature, databases of similar equipments, calculations based on machine parameters or, sometimes, try to find similar equipments running and just going there and measure it.

In any case, an important decision to be taken is on how to model each noise source. There are three basic types of sources which one can introduce in the model:

- Point sources – Adequate for small sources, such as fans, or larger sources, with well-balanced dimensions, sitting away from relevant receivers;

- Line sources – Adequate for linear shaped sources, such has piping, conveyors, as well as moving sources paths;

- Area sources – Can be vertical, such as openings in a building, noise radiating façades, or very large machines, or horizontal, such as a roof, or a number of fixed or moving sources distributed on the ground.

Examples of sources modelled as (from left to right): point sources – fans on top of deposits, chimneys; line sources – high pressure pellets transportation pipes; plane sources – light metal façades and roof, openings.

The way calculations are made by modelling software such as CadnaA, can be set up to comply with different methods and standards. The most common for industrial sources is the method of ISO 9613, which details we will not go into here. One must also configure correctly a number of software parameters, related with the calculation configuration, such as the maximum search radius, minimum distance source to receiver, maximum reflection order, minimum distance receiver-reflector, reference time, with the grid calculation (for noise maps), such as receiver spacing and receiver height, or with each source in particular, such as single band or spectrum, directivity, geometry, and type of noise levels input to the source (e.g. L_p measured at a certain distance outdoors, L_i impinging on a façade from the interior side, etc).

The hard work involved on building and validating the acoustical model of a large industrial plant is highly compensated when one, finally, gets the final model to produce results which make sense and match the reality. It is then that one can start taking advantage of the model for practical applications, such as source ranking, calculation of individual source contributions to the total noise at any given receiving point, evaluation of different scenarios of noise control actions to propose an action plan and, of course, fully running the model to get noise maps, creating calculation grids which one can even use for further grid operations, such as arithmetic or logarithmic addition or subtraction.

Practical Examples of Application

Chemical Plant

This noise mapping project came from the need of the company to renew its IPPC Environmental License, having indications that it was not fully complying with noise limits imposed by the Portuguese regulations. Therefore, a comprehensive noise source survey was carried out, estimating its octave band sound power level from sound pressure level measurements close to each source. As happens with most process industries, most relevant sources are located outdoors and they run 24 h a day. Input data has been appropriately validated, according to the methodology described above: source validation followed by model validation – this was not particularly difficult in this case, due to the fact that the plant is located in the countryside, away from other noise sources, except for a national road, but with sparse traffic. However helpful this may be for model validation purposes, the fact that background noise is very low does not help when it comes to noise control requirements as, according to regulations, one must reduce plant noise down to the background level. The present regulations, however, do imply that, in case background level is lower than 42 dB(A), which was the case at most sensitive receivers in this project, the limit for particular noise from the factory, at any time, is also 42 dB(A), including the correction factors for tonal components (+ 3 dB(A)) and for impulsive noise (+ 3 dB(A)). From the acoustical model, a source ranking has been performed, identifying the sources which generate more noise to the sensitive receivers, which in this case are just a few isolated houses, located to the opposite side of the national road.

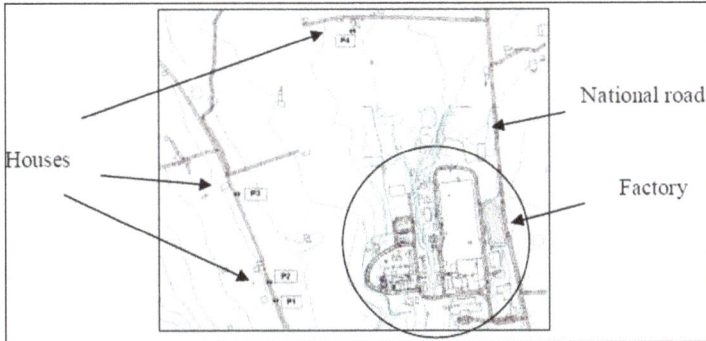

Location of the factory, road, houses and corresponding measurement points P1, P2, P3, P4.

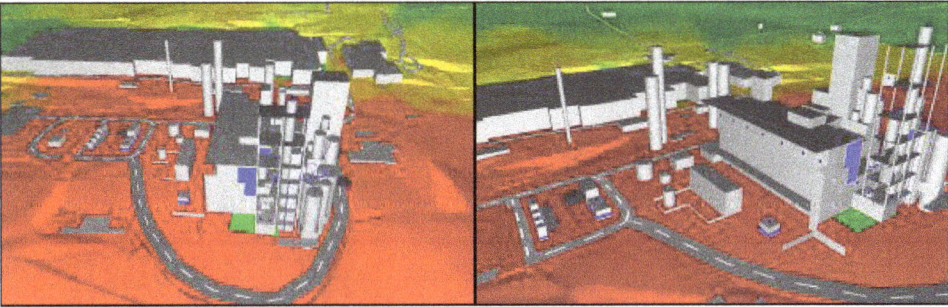

3D drawings of the acoustical model, with projection of the noise ma pinto the terrain.

Food Industry

This example is almost the opposite of the chemical plant, as it is a traditional old plant, located right in the middle of the city, with residential buildings all around, the background noise levels are high and most noise sources are inside buildings, except for chimneys. Although the factory does not actually stop at night, some sections of it do stop, therefore reducing its activity, and noise generation, during night time. The project started due to a complaint from a neighbour leaving near the factory. The project consisted of developing the acoustical model, produce noise maps of present situation to assess the noise impact and communicate it to the Municipality, specify a noise control action plan, implement it and measure the final results, updating the noise map in the end.

Two types of sources were identified as relevant to the noise emission: interior sources, which radiate noise to the outside through the vast number of windows, and chimneys, located above the roofs. The first were not taken individually – the approach has been to measure noise impinging on the windows from the inner side, and use the CadnaA feature "L_i from interior sources" together with the transmission loss of the windows to calculate the sound power lever per unit area radiated to the outside by vertical area sources, which were used to model the windows. The latter were measured and inserted in the model one by one as chimney sources, with a frequency dependent directivity, simulated by CadnaA from the known velocity and temperature of the gas flow at each chimney.

In this case, apart from the regular validation process, attention has been focused at the most critical receivers, namely at the house of the complainant, out of which window the microphone was mounted. Continuous long term measurements were taken, for several days including week and weekend days, using a PC-based noise analyser, with audio recording. This helped in identifying noise sources, also enabling the filtering out of background noise events such as car pass-by during the more silent periods, such as night and weekend. Both the model and the measurements agreed that the number one sources were noisiest chimneys located above the building closer to the complainant house, followed by windows of the noisiest floors of the factory buildings with façades directed towards the afore mentioned house.

An action plan was specified in two steps, the first of which has been already implemented. Control measurements have been taken, including another long term measurement at the complainant window, which show a clear noise reduction, within the expected from the model.

3D view of the CadnaA model and corresponding photo, both views taken from the window of the complainant's house.

Noise maps for noise indicators L_n (left) and L_{den} (right).

Foundry

This Project consisted on the production of Noise Maps in the context of an Environmental Impact Study for the expansion of the Foundry, which needed to increase its production capacity, for a number of scenarios: present situation, future situation with no noise abatement measures and future situation with noise abatement measures. The specification of these noise abatement measures was also part of the project. The present situation noise map had already been produced in the past, although it had to be updated to the new Portuguese noise law, and it was complying to the regulations. As the foundry was going to enlarge and new equipments were to be installed, the aim of the company was to study the problem in order to guarantee that it would still comply with the noise limits after the expansion project has been concluded.

A complete survey was carried out of all the changes which would take place, including new equipments, new layouts and new buildings. In this case, sound power levels from most new equipments were available and were introduced in the model. Noise levels at critical points were calculated and source ranking was made in order to identify which noise sources needed to have special noise control conditioning measures prior to its installation, which could include relocation relative to the initially planned, in order to maintain full compliance with the noise limits. Next figures depict noise maps without and with the noise control measures as well as a map of differences, obtained by grid subtraction of the noise maps of the two scenarios.

Noise maps of the foundry after the expansion without (left) and with (right) noise control measures.

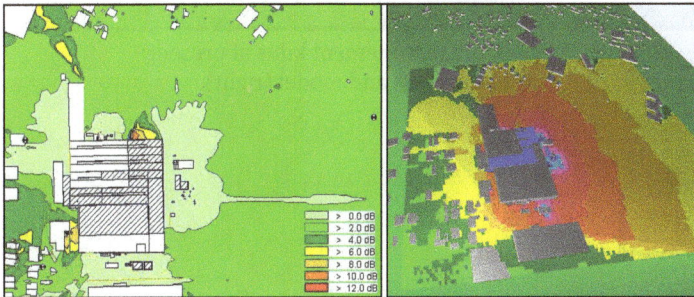

Map of differences (without minus with noise control measures) and 3D visualization of the noise map with the noise control measures.

Cement Industry

This example relates to a large cement factory which was starting the implementation of a large investment plan on its production lines which consisted basically on the replacement of three existing lines, which were old and ineffective, by a new production line, with a new kiln, with higher production capacity. Due to lack of space and presence of an urban agglomeration nearby, it was a complex operation and noise was one of the major issues as it should comply with ever more stringent regulations. Therefore, a noise mapping project has been carried out, which enabled the simulation of three main stages: present situation, transition situation and final situation with the new production line. It was shown that, at present, the factory is not complying with legislation and, although a noise reduction is predicted with the new layout of the factory, it has been shown that there is a high risk of not complying with the regulations. In this context, a noise control plan was studied, aiming at the full compliance with noise limits when the project is finished, i.e., at the final stage with a single production line.

Noise map of the present situation (left) and of the future situation with noise control measures (right).

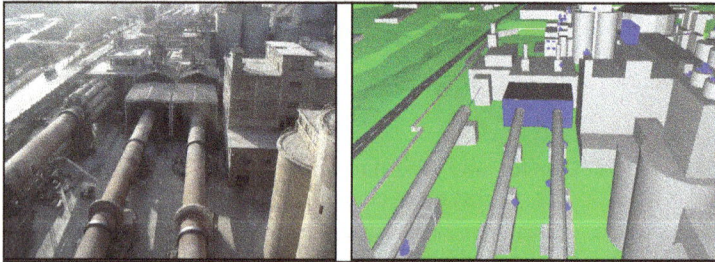

View of the three cement kilns at present:
photo (left) and model (right).

Port Noise

When you think about a seaport, noise is hardly the first thing that comes to your mind. The very first ideas are probably more of a visual character: vessels berthing, passengers arriving, big cranes loading and unloading vessels lying by the quay. There is activity everywhere; a port should not be still. All this activity produces, of course, some sounds. A completely quiet port is nothing to strive for.

But, how does a port actually sound? Should you anticipate the sounds of the sea like the buzzing wind, waves washing against the pier, accompanied with some seagull cries? Or are you about to hear deafening jar from the working machinery, pounding bass tunes from the enormous diesel engines and whining and rattling from the unloading bulk vessels? The reality might roam somewhere between these two extremes, and we can state some questions like: how should a port sound like? How is a good port sound environment like? And why are these questions important to ask in the first place?

The noise question is much more complex than just decibel levels and acoustic measurements, which are already a whole science of their own. When you start to investigate noise, you will be cast onto different juridical, economical, medical, psychological, ecological, technical, architectural, social and societal as well as aesthetical issues which all can be relevant even from a port's perspective.

The image of sea ports has changed remarkably during the past few decades. In the old days, the word dockland could evoke quite dubious associations, and ports were often located outside the inner-city areas. Now, many port cities have grown so that ports have been encircled by the city settlements, often with residential areas close-by. The ports themselves have been modernized and become efficient logistic hubs with less manpower and more technology. This has been a big change, also from the image point of view. In many European port cities, flashy and dynamic inner-city residential areas are being built close to the port areas, and the main target group for these housing projects is the economically privileged upper middle-class. The dominant city planning philosophy is to mix different functions in the same area, and this can, at least in some cases, lead to conflict of interest. The city planners' visions don't meet the ports' reality entirely smoothly in all cases.

When it comes to legislation regarding noise and other environmental impacts, the ports are handled the same way as industrial establishments in all the three countries. In many respects, this is justified due to the several parallels ports have with different industries. Alike many industries, there are multiple environmental impacts and several emission sources in a port. But at the same time, this leads to more strict noise regulations for ports than other modes of transport. Of the studied countries, the juridical situation is most complex in Sweden, and many of the questions regarding the ports' condition remain unanswered. When we discuss port noise, it is important to keep in mind that ports in general are not particularly noisy places compared with an average city street where the sounds of car and truck traffic dominate the soundscape. The port noise is only a little addition to the sound carpet that embraces the city from every direction.

Even though ports stand only for a little fraction of all environmental noise our society, the noise issue cannot be ignored by ports. The noise abatement requirements from the environmental authorities will not be less strict in the future, and at the same time, minimizing the environmental impact and being a good neighbour will be crucial for the ports' social responsibility and public relations. Due to the rising environmental awareness, all companies, and therefore ports as well, will be scrutinized even more carefully by the authorities, the media and the public in the future. That is why a systematic noise abatement work is an important issue for the ports' future development.

Noise Sources in the Port Environment

The significance of the noise question varies greatly from port to port. Both the location of the port, its topography and the characteristics of the port operations influence the noise situation greatly. Measured in decibels, ports dominated by cargo traffic are typically noisier than ports dominated by passenger traffic. However, passenger-oriented ports located in inner-cities are more often struggling with the noise issue. This is simply due to the fact that cargo-oriented ports are usually located further away from residential areas.

A port environment has typically several noise sources. The presence and the significance of the sources vary depending on the type of traffic in the port. The following noise sources are the most common ones:

- Working machines: Cargo handling equipment is a significant noise source in ports. Examples of these machines are cranes, reach stackers, straddle carriers, terminal tractors and trucks. Noise sources on these are engines, exhaust systems, tyres and alarm signals. The cargo handling equipment is usually driven by diesel engines, and noise is generated both by driving and cargo handling events. The engine solutions are more or less the same as the ones used for road vehicles, but unlike road vehicles, noise reducing solutions for this kind of machines have not been required by legislation. Nevertheless, newer equipment is generally more silent than older. Some of the improvements have been achieved through a conscious development work, but sometimes improvements have been resulted by other requirements. An example of this is the requirement of catalytic converters in the exhaust systems of the terminal tractors, which has reduced the noise levels as a biproduct.

- In and out truck and car traffic: As ports are hubs where different traffic modes meet, the in and out road traffic to ports is extensive. This creates problems both with air pollution and noise in the port areas and their vicinities. Even passenger vessels generate car traffic to the ports, because many passengers arrive to the ports by car or have their cars with them during the boat trip on RoPax vessels typical to PENTA ports. Creating functional road and street traffic solutions and eliminating the nuisances caused by this traffic at ports is a big challenge to both the ports and the urban planning. For ports dominated by heavy truck traffic this is even a bigger challenge.

- Railway: The ports with a railway connection have special noise questions related to the rail operations to handle. The most problematic of these sounds is the impulse noise generated when the railway wagons are shunted.

- Vessel-quay interface ramps: The ramps between the vessels and the quay which vehicles enter and leave the vessels are usually made of concrete and metal. Driving on the ramps can create loud impulse sounds.

- Cargo handling: Containers, bulk cargo - Cargo handling sounds in ports are many and diverse. Generally, unitised traffic generates less noise than loading and loading bulk cargo. This noted, even container handling creates impulse noise when the containers are dropped to the ground, to the vehicle or are clashing into each other. Unloading liquid bulk can create tonal, whining sounds. Maybe the most problematic noises in the port environment are created by loading and unloading some types of dry bulk, which creates loud impulse sounds that are quite difficult to abate.

- Vessels: Last but not least, the vessels themselves are maybe the most significant noise source in ports. There are several noise sources on vessels: engines, auxiliary engines and their funnels and exhaust systems, different kind of ventilation and air conditioning systems, hydraulics, pumps and on-board ramps. For several reasons, the vessel noise is a real dilemma.

Ramps are a significant noise source in the vessel-quay interface.

Within the EU project efforts acoustical properties of the different noise sources in the port environment were measured and psychoacoustic descriptors were used to approximate the annoyance caused by the different noise sources. In the study, the sounds with highest decibel levels were not the same as the most annoying ones. The most annoying port sounds were alarm signals, gantry cranes and vessel exhaust stack ventilation.

Technical Challenges

The noise abatement efforts in ports are constrained by several challenges, both of technical, legal, financial and organisational character. The most important technical challenges are:

- Outdoors environment: Port operations take usually place in a wide outdoors area, where noise easily can spread to neighbouring areas. Abatement measures, such as noise walls and barriers are not always possible to build due to for instance lack of space. Especially propagation towards water is difficult to hinder with this kind of methods. The weather, as wind, temperature and air pressure, can also greatly impact the propagation of noise in outdoors environment. In addition, the outdoors environment can complicate the noise measurements and make it difficult to determine if a noise disturbance is caused by the port or some other source. In city areas, the background noise levels are high, and the proportion of the port to the overall noise levels can be difficult to show.

- Acoustically hard materials: Acoustically, there are hard and soft materials. Generally, hard materials conduct sounds, and soft materials muffle them. In port environment, hard materials such as concrete, asphalt and metal surfaces are common.

- Closeness to water: Water, which ports are by nature surrounded by, is the acoustically hardest of all materials. It conducts noise easily to opposite shores and nearby islands. Isolating the port from the surrounding areas by noise walls or barriers is complicated, and often impossible.

- Several different noise sources: As port noise consists of sounds emitting from several different sources, to cut down one of the sources does not necessarily have any impact on the overall noise level.

- Scattered noise sources on different heights: Noise from different positions in the area complicates mitigation measures. Propagation from sources on the ground level, like tyres, is easier to hinder than from sources higher up, such as the funnels of the vessels. Moreover, some of the noise sources, as working machines and vessels are moving, which makes stationary mitigation measures inefficient.

- Low-frequent, tonal and impulse noise: Compared to average traffic noise, which is quite monotonic in character, port noise is, due to the different sources, more diverse. The acoustic properties of port noise make it more annoying to hear and more complicated to abate. Low-frequent noise is typical for vessel engines, and to muffle it requires big silencers or thick noise barriers. Impulse noise is typical for cargo handling operations and driving on vessel-quay ramps. Noise with tonal elements is typically emitted from fans of the vessels and beacons of the working machines. Common to these three types of noise is that they all are experienced as more annoying than the average traffic noise.

- Best Available Technology principle not generally applicable. The location and the nature of operations are of great importance for a port's noise situation.

Together with the topography and the layout of the port area these factors make every port unique when it comes to the noise propagation situation. In environmental justice, the principle of Best Available Technology (BAT) is widely used to benchmark the best technical standards to prevent hazards to the environment. The operators are required to use BAT whenever it is economically feasible. Since the variation between ports regarding the noise question is great, all noise abatement measures have to be tailor-made to fit each port. Therefore, the BAT principle cannot be directly applied to port noise.

The Dilemma of Vessel Noise

Vessels are in many ways the most challenging noise source in ports. Firstly, the technical and acoustical features of vessel noise make it problematic as such. Vessels are, as a rule, running their auxiliary engines to produce electricity they need during the time they are berthed. The sound from the engines is low-frequent, which makes it more annoying to hear. Low-frequent noise has a long wave length, and this means that muffling it requires big, space-consuming silencers on the vessel. If the noise is not muffled on the vessel, standard noise walls, soundproof windows and like are insufficient to mitigate it from penetrating the nearby buildings.

Noise emitting from vessels is not regulated internationally. For International Maritime Organization, IMO, vessel noise is primarily an occupational health question, and they are also working on recommendations for noise emitting to the water to protect the marine fauna. Noise emissions to the air are not on agenda for IMO at the moment, so there is no regulation in sight in the nearest future. Therefore, noise emissions from maritime traffic are only regulated on the national level through the environmental permits of ports.

As port noise is classified as industrial noise, the noise emissions from a vessel become industrial noise as soon the vessel enters the water area of the port. From the ports point of view, the situation is very problematic. The maritime traffic is per definition international, but the noise requirements vary from country to country and from port to port. In this situation, it is not all too simple to implement requirements. A strict noise policy is also seen as a competitive disadvantage by the ports.

Finding solutions to the dilemma of vessel noise requires a good collaboration between the ports and the ship owners. In liner-traffic, long-term customer relationships between the ports and the ship owners are created, and it is uncomplicated to bring the environmental issues, such as noise, to the agenda. The situation is more challenging with irregular customers such as cruising vessels which may only visit a port once or twice a season or cargo vessels using the port occasionally. In the cargo segment, the transport and logistics chains are so complex that the transport buyers usually do not even know how the goods are transported. The vessel noise is a quite abstract question to a transport buyer who, if environmentally oriented, is more interested of carbon dioxide emissions and the climate impact of the transport.

In the passenger segment, the general environmental consciousness creates some customer pressure to environmental friendly solutions. In addition, a vessel with low external noise emissions has also a higher customer comfort. Therefore, incentives to find silent solutions exist in a way which is non-existent in the cargo segment.

Wind Farm Noise

Wind turbines generate two types of noise: aerodynamic and mechanical. A turbine's sound power is the combined power of both. Aerodynamic noise is generated by the blades passing through the air. The power of aerodynamic noise is related to the ratio of the blade tip speed to wind speed. The mechanical noise is associated with the relevant motion between the various parts inside the nacelle. The compartments move or rotate in order to convert kinetic energy to electricity with the expense of generating sound waves and vibration which is transmitted through the structural parts of the turbine.

Depending on the turbine model and the wind speed, the aerodynamic noise may seem like buzzing, whooshing, pulsing, and even sizzling. Downwind turbines with their blades downwind of the tower cause impulsive noise which can travel far and become very annoying for people. The low frequency noise generated from a wind turbine is primarily the result of the interaction of the aerodynamic lift on the blades and the atmospheric turbulence in the wind. High frequency noise is also generated due to the interaction of the air turbulence and the blades during their rotation, constantly changing angle of attack.

Depending on the rotational speed of the turbine, the size and the airflow wind turbines emit infrasound, low and high frequency sound waves. Large wind turbines produce infrasound of 8–12 Hz range. Small turbines can achieve higher blade tip velocities that can give low frequency noise of 20–20 kHz range.

The levels of infrasound radiated by the large wind turbines are very low in comparison to other sources of acoustic energy in this frequency range. However, the annoyance is often connected with the periodic nature of the emitted sounds rather than the frequency of the acoustic energy. Because low frequencies travel farther than the high frequencies due to their long wavelengths, they become a cause of irritation for residents living not so close to wind farms.

Sound is a series of waves that travel through air in the form of disturbances and reaches our eardrums. Any natural or artificial obstacles such as hills and buildings play a shielding effect role, reflecting the sound, while most of the times absorbing some of the acoustic energy. Trees and ground vegetation also attenuate sound and change its directivity patterns. The distance and obstacles that are located between the source and the receiver have a significant impact on the acoustic 'line of sight'.

The ways to reduce noise from wind turbines are mostly focused on blade design optimisation and choice of wind farm location as it still a new field and new techniques are under development. A major challenge that needs to be addressed before abatement techniques are put in action; is to establish accurate measurement methods of wind turbine noise in order to define the parameters of the problem and find an appropriate acoustic solution for it. Noise measurements from wind turbines is a complex task because the background noise levels are comparable to the noise levels from wind turbine when it starts operating at certain wind speeds. This is the reason a commonly used approach that overcomes the above issue needs to be established.

Current acoustic treatment methods range from designing quieter wind turbine compartments to placing wind turbines as far as possible from residential areas. Wind turbine blades are aerodynamically shaped to avoid causing abrupt air flow disturbances and the turbine is placed upwind to eliminate interference of the flow between the tower and the wind turbine blades. Apart from aerodynamic solutions other treatments include sound proofing of the nacelle to reduce mechanical noise and careful selection of the wind turbine site to attenuate noise until it reaches the receiver.

Table: Important frequency ranges.

Description	Frequency range
Normal hearing	20 Hz to 20 kHz
Normal speech	100 Hz to 3 kHz
Low frequency	20–200 Hz
Infra sound	<16Hz

It is common sense that the extensive use of wind turbines is included in the current and the future environmental plans of every country, fact which will certainly place a strong demand for dealing noise issues in the near future.

Noise Types and Patterns

Sources of Wind Turbine Sound

The sources of sounds emitted from operating wind turbines can be divided into two categories: (1) mechanical sounds, from the interaction of turbine components, and (2) aerodynamic sounds, produced by the flow of air over the blades.

More specifically, wind turbines produce energy by the rotational motion of the blades due to the wind flow. Rotating blades are known to emit three different types of acoustic signature:

- Tonal noise,

- Broadband noise,

- • Mechanical noise.

Tonal noise is characterised by discrete frequencies and it is generated by the periodical rotation of the turbine blades. Tonal noise is caused by the unsteady air velocity due to the blade rotation which disturbs the flow on the blade surface. It is directional as it is produced at the direction the blades meet the airflow and therefore is dependent on the observer position.

As the blades rotate the load on the blade surface changes periodically causing analogous changes in the unsteady pressure on the blade surface inducing sound waves. Depending on the position of the observer in relation to the turbine while the blade is in operation the observer receives variable acoustic signals. Also the sound component in the direction of the observer varies with time and a sound wave is generated. Normally, the flow through the blades is distorted (nonuniform) and therefore the angle of attack of each blade varies continuously as the turbine rotates causing the sound generation to be highly directional and more frequent. This change in the angle of attack can be very abrupt especially when velocity discontinuities occur in the inflow profile resulting in rapid changes in the blade loading, flow disturbance and therefore generation of acoustic waves.

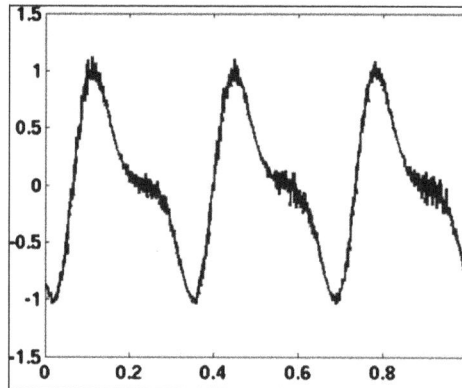

Tonal noise time history.

Such type of acoustic waves can be produced when the mounting tower of the wind turbine interferes with the flow passing through the wind turbine blades. This causes local velocity instabilities in the flow field which moves through the blades producing pulsing low frequency noise. It becomes self-explanatory that when the reverse order occurs, the disturbances ease as the blades encounter only the free field flow disturbances.

The noise generated by downwind designs has low frequency in the range of 20–100 Hz and it is caused when the turbine blade encounters localised flow deficiencies due to the flow around a tower. It can also be impulsive described by short acoustic impulses or thumping sounds that vary in amplitude with time. It is caused by the interaction of wind turbine blades with disturbed air flow around the tower of a downwind machine.

Infrasound

A special category of the tonal noise released by wind turbines is infrasound. As we have already mentioned while low frequency ranges at the bottom of human perception (10–200 Hz), the infrasound is below the common limit of human perception. Sound with frequency below 20 Hz is generally considered infrasound, even though there may be some human perception in that range. A distinctive characteristic of infrasound is that it can travel very far because of its long wavelength that dissipates with a low rate and therefore it makes it easier to 'survive' and be present in our everyday life.

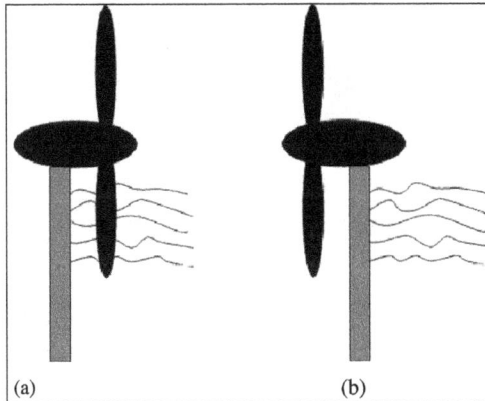

(a) Downwind turbine; (b) upwind turbine.

Considering the nature of infrasound, it becomes apparent that downwind wind turbines are more likely to produce significant levels of infrasound levels. This is because the tower–blade flow interaction creates air velocity instabilities that produce very low frequency acoustic waves called infrasound. Although downwind wind turbines have been used in the past, the modern technology has moved away from that noisy design to the upwind designs that give higher frequency noise levels and therefore are less irritating for humans.

Broadband noise is dominated by high frequencies greater than 100 Hz and it is characterised by non-periodic signals that constitute an envelope that varies periodically. Its main source of generation is the interaction of the wind turbine blades with atmospheric turbulence, and also described as a characteristic "swishing" or "whooshing" sound.

To determine the relative importance of tone noise and broadband noise we consider the narrow-band frequency spectrum of a signal. At higher frequencies the broadband random noise dominates the spectrum.

Mechanical Generation of Sound

A wind turbine consists of mechanical components that move or rotate in order to capture the motion of the turbine and convert it to energy. Sources of such sounds include:

- Gearbox,

- Generator,

- Yaw drives,

- Cooling fans,

- Auxiliary equipment (e.g. hydraulics).

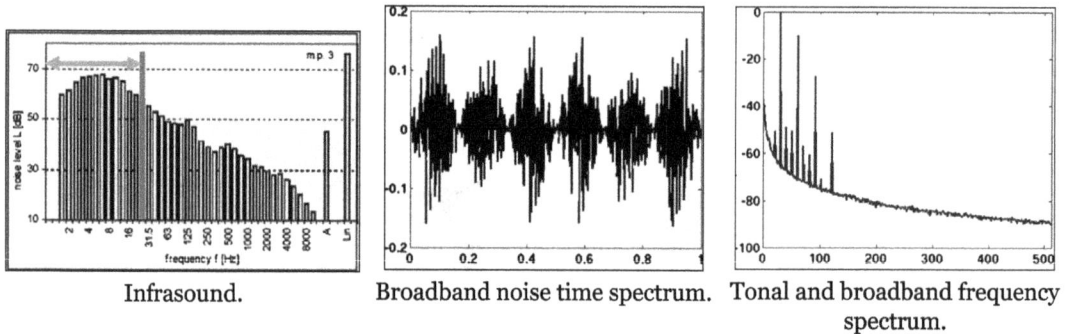

Infrasound.   Broadband noise time spectrum.   Tonal and broadband frequency
                                                                                spectrum.

Generally the mechanical sound is low frequency sound although it might have a broadband component that comes from the relative motion from each of the above parts. The turbine's metal parts come in contact with each other, such as the generator, the gearbox and the shafts and they emit noise and as they vibrate.

Because wind turbines can have different constructions they might have different sound emissions because of the way in which they operate. For instance they may have blades which are rigidly attached to the hub or may have blades that can be pitched (rotated around their long axis). Some have rotors that always turn at a constant or near-constant speed while other designs might change the rotor speed as the wind changes. Wind turbine rotors may be upwind or downwind of the tower.

It is worth mentioning that the hub, rotor, and tower may act as loudspeakers, transmitting the mechanical sound and radiating it. The transmission of noise from the mechanical parts of a wind turbine can take place in two ways:

- Structure-borne,

- Air-borne.

Structure-borne sound is a sound that is propagated through structures as vibration and subsequently radiated as sound. The intensity and the frequency of structure-borne sound depend on many factors such as the rotational speed of the wind turbine, as well as the type and the material of the mechanical parts that vibrate. Air-borne means that the sound is directly propagated from the component surface or interior into the air.

Structure-borne sound is transmitted along other structural components before it is radiated into the air. For example, type of transmission path and the sound power levels for the individual components for a 2 MW wind turbine. Note that the main source of mechanical sounds in this example is the gearbox, which radiates sounds from the nacelle surfaces and the machinery enclosure.

Utility scale turbines are usually insulated to prevent mechanical noise from proliferating outside the nacelle or tower. Small turbines are more likely to produce noticeable mechanical noise because of insufficient insulation.

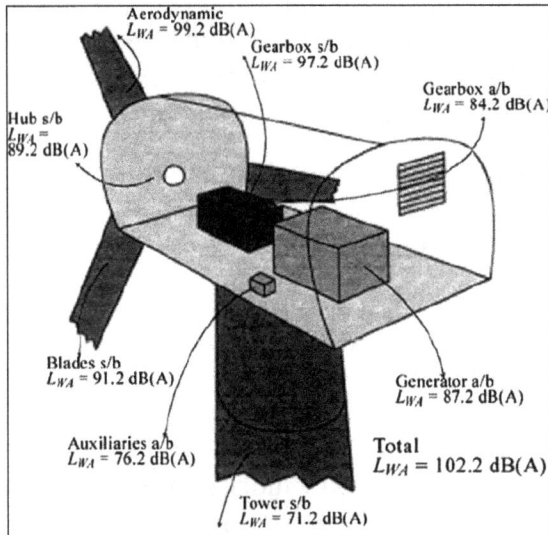

Sound power levels of wind turbine components.

Factors that Affect Wind Turbine Noise Propagation

Propagation refers to how sound travels. Attenuation refers to how sound is reduced by various factors. Many factors contribute to how sound propagates and is attenuated, including air temperature, humidity, barriers, reflections, and ground surface.

The ability to hear a wind turbine also depends on the ambient sound level. When the background sounds and wind turbine sounds are of the same magnitude, the wind turbine sound gets lost in the background. The most important factors are:

- Source characteristics (directivity, height),

- Distance of the source from the observer,

- Air absorption,

- Ground effects (reflection and absorption on the ground),

- Weather effects (wind speed, temperature, humidity),

- Shape of the land – land topology.

Source Characteristics

The source characteristics such as height and directivity can affect the sound propagation path and its power or intensity. The higher a source is located, the higher the sound power loss rate is. This means that wind turbine that is mounted on a tower relatively high to a residential estate it has relatively low noise impact on the residents as the sound energy attenuates until it reaches the human ear. The directivity of an acoustic source has also a significant impact on the sound perceived by the human ear. For example when the sound is forced to follow a certain directional path determined by the geometrical shape it is placed in, such as conical speaker, the radiation field is concentrated towards a certain area leaving quite zones in the opposite direction.

In general, as sound propagates without obstruction from a point source, the initial sound energy decreases and it is being distributed over a larger and larger area as the distance from the source increases.

For example, in the case of spherical excitation or a monopole noise source the sound is radiated in all directions and the sound level is reduced by 6 dB for each doubling of distance from the source.

A moving train is a line source and it emits equal sound power output per unit length of the train line. A line source produces cylindrical spreading, resulting in a sound level reduction of 3 dB per doubling of distance. The spherical propagation is associated with the three-dimensional propagation and the line source with the one-dimensional sound wave propagation, respectively. When two monopole sources of equal strength but opposite phase are put together at a short distance they produce a dipole, which is referred as two-dimensional sound wave propagation.

Air Absorption

Air absorption of sound is driven by two mechanisms: molecular relaxation and air viscosity. Molecular relaxation is the transition of a molecule from an excited energy level to another excited level of lower energy. High frequencies are absorbed more than low because they have short wave length and therefore the waves dissipate as they travel through the air molecules. The air absorption must be taken into account at high frequencies when calculating the reverberation time of a room. It is due to friction between air particles as the sound wave travels through the air. The amount of absorption depends on the temperature and humidity of the atmosphere.

Ground Absorption

The ground can contribute to the sound attenuation by two mechanisms sound absorption and sound reflection. When the sound hits the ground the acoustic energy loss depends on the reflection coefficient of the surface. On hard surfaces attenuation occurs due to the acoustic energy losses on reflection while on porous surfaces, sound levels are being reduced due to the increased absorption of the ground. High frequencies are generally attenuated more than low frequencies.

The reflection coefficient depends on the impedance of the two media, in this case, air and ground, and represents the absorbency of the ground in a homogeneous

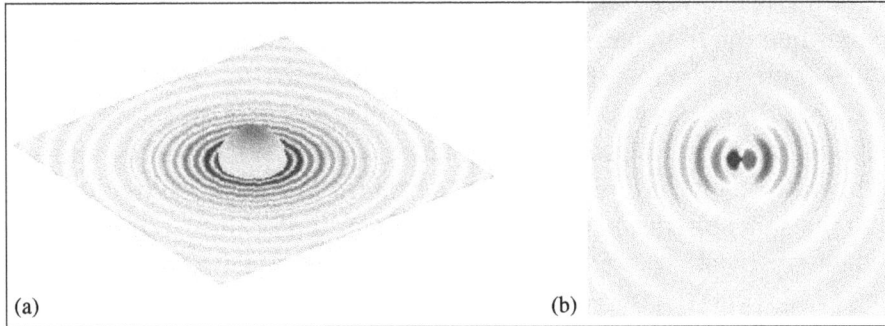

(a) Monopole; (b) dipole.

When the source and receiver are both close to the ground, the sound wave reflected from the ground may interfere destructively with the direct wave. This effect (called the ground effect) is normally noticed over distances of several meters and more, and in the frequency range of 200–600 Hz.

Land Topology

The topology and the shape of the land can significantly affect the magnitude and the direction of sound. For example, trees and high altitude vegetation can contribute to the sound attenuation. However, a long series of trees several hundreds of meters long is required in order to achieve significant attenuation.

Also significant attenuation can be achieved by the use of natural or artificial barriers or obstacles such as hills and buildings that exist on the ground. The level of impact on the sound reduction of an obstacle depends on whether it is high enough to obscure the 'line of sight' between the noise source and receiver.

Due to their short wavelength, high frequencies are trapped by the obstacles preventing them from travelling far, unlike the low frequencies.

Similarly to the ground reflection theory, the material of the barriers plays a dominant role in the sound propagation and this is the reason barriers are often used for noise treatment purposes. A barrier is most effective when placed either very close to the source or to the receiver.

Weather Effects, Wind and Temperature Gradients

The wind and the temperature can affect the propagation of the sound in the atmosphere under certain weather conditions. The mean uniform wind flow determines the background noise levels and it alters the sound pressure downwind and upwind. When a wind is blowing there will always be a wind gradient. A wind gradient results in sound waves propagating upwind being 'bent' upwards and those propagating downwind being 'bent' downwards.

The temperature is another factor that affects sound radiation however; it becomes important only when the high temperature gradients occur. Such dramatic changes in the temperature profile are unlikely to happen in the atmosphere close to the ground but they can occur at high altitude layers. Any temperature differences in the atmosphere can cause local variations in the sound speed since the latter depends on the temperature of the gas. Higher temperatures produce higher speeds of sound. When sound waves are propagating through the atmosphere and meet a region of non-uniformity, some of their energy is re-directed into many other directions. This phenomenon is called refraction.

Measurement Techniques and Challenges

Low frequency noise emissions from wind turbines have given rise to health effects to neighbours. Resident complaints about the irritating noise from the wind turbines has led scientists and engineers to invent ways to assess the levels of noise in the near and in the far field and eventually close to dwellings where residents live.

Measuring noise from wind turbines is not an easy task due to the fact that background noise levels increase as the wind speed increases. So, as the wind turbines start rotating the background noise levels are being intensified. Also when we think that the wind turbines are placed outdoors where trees, leaves and vegetation in general, are present then it becomes apparent that the background noise is comparable to the noise emissions from the turbines. This makes it extremely difficult to measure sound from wind turbines accurately. At wind speeds around 8 m/s and above, it generally becomes a quite abstruse issue to discuss sound emissions from modern wind turbines, since background noise will generally mask any turbine noise completely.

To assess the potential levels of infrasound and low frequency noise around a wind farm and at neighbouring locations of interest, the measurements are undertaken using a measuring system capable of capturing frequencies from 1 Hz to 20 kHz. Measurements are performed at internal and external locations placing the microphones at locations where noise is considered more audible when occurred. Assuming that the appropriate equipment is being used and the calibration procedures have been followed, the standard procedure to take noise measurements is the following:

- Ambient background noise levels evaluation.

- Spot measurements of noise levels inside dwellings subject to access or prediction if access is prohibited.

- Exterior acoustic measurements at neighbouring facades to assess annoyance.

- Spot measurements of wind speed.

- The evaluation and reporting of measurements made.

It is worth mentioning that there is a variety of sophisticated model available but they are all at developing stages.

For Small Wind Turbines

Our experience indicates that in practice, field measuring is a challenging task not only due to the difficulty to estimate background levels but also due to the complexity of the required experimental set up. One has to count for the directional noise emission depending on which side the wind is blowing and the fact that modern small wind turbines rotate around their vertical axis making the noise measurement techniques even more demanding.

Because of the importance of background noise in determining the acceptability of the overall noise level, it is crucial to measure the background ambient noise levels for all the wind conditions in which the wind turbine will be operating. Sound propagation is a function of the source sound characteristics (direction and height), distance, air absorption, reflection and absorption and weather effects such as changes of wind speed and temperature with height.

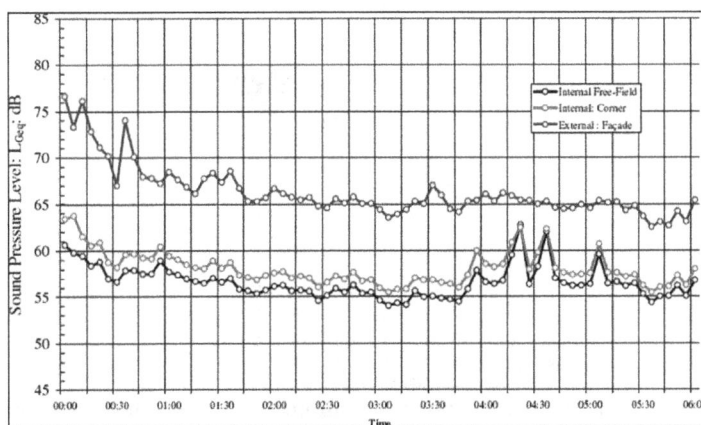

Time history of measured sound pressure levels.

Given the current situation difficulties, a good idea to tackle with such a problem would be at a first stage to use the anechoic chamber to remove background noise and depending on the results, to employ measurement techniques and methods which enable characterisation of the noise emission from wind turbines at a receptor location. The use of a vertical board mounted with a microphone suggested in the document of IEA recommended practices in case of such a problem. The document recommends the use of a large vertical board with a microphone at the designated position with its diaphragm flushed with the board surface or to suppress wind induced noise on a microphone.

Abatement Methods

In principle there are two ways to reduce noise: either make the source less noisy or make the receiver more sound proof. The same principle can be applied to the wind turbines when considered as noise sources.

Also if we think the factors that affect the noise propagation we can easily guess a few measures we can take in order to control noise from wind turbines. More specifically locating wind farms as far and as high as possible from residential areas and improving the sound insulation in houses we can significantly limit the sound levels received by the human ears. Natural obstacles and vegetation can also prevent noise from reaching people's homes.

A systematic and scientific consideration of all those factors along with real time measurements leads to a study known as environmental impact assessment. Such assessment is essential to evaluate the current environmental conditions at a proposed erection of a wind farm. Among others, it assesses how the proposed wind farm will affect the environment and the public health, and whether any resultant changes in conditions are deemed acceptable or unacceptable. Best code guidelines and policy standards are being practised and followed in order to ensure environmental protection including noise and nuisance.

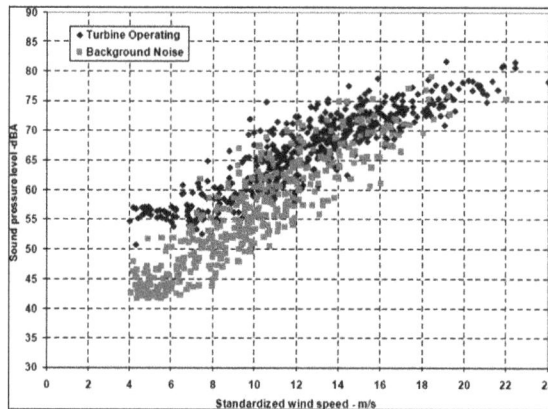

Background and operation noise measurements for a small wind turbine.

Recalling that the type of noise generated from a wind turbine can be mechanical and aerodynamic, we are able to investigate and find solutions for noise treatment.

Starting with the mechanical parts of a wind turbine that rotate or move in relation to each other such as gears that are creating structure-borne noise and vibration, it becomes clear about the need for designing gearboxes for quiet operation. Wind turbines use special gearboxes, in which the gear wheels are designed slightly flexible in order to reduce mechanical noise. One way of doing this is to ensure that the steel wheels of the gearbox have a semi-soft, flexible core, but a hard surface to ensure strength and long-time wear.

In addition to designing quiet parts, insulating those parts seems to be another way to tackle that type of noise. Soundproofing and mounting equipment on sound dampening buffer pads helps to deal with this issue. In addition, special sound dampening buffer pads separate the gearboxes from the nacelle frame to minimise transmission of vibrations to the tower.

There are limited acoustic solutions that can be applied to reduce noise from the mechanical movements due to the space restriction inside the nacelle as well as the necessity not to disturb the efficient functionality of the mechanical parts with added acoustic treatment. Although for large wind turbines mechanical noise is not much of an issue since those are located far from dwellings, for the small house, mounted wind turbines vibration can have significant impact on people's lives living in the house where the wind turbine is attached.

Moving on to the aerodynamic noise, there are various techniques or even technologies to decrease sound from the wind turbine blades. As somebody would expect most of those techniques originate from designing more aerodynamic blades and adjusting the rotational speed of the turbine.

Some of the noise causes we have discussed concern downwind designs, blade speed and shape and interaction of the airflow between the tower and the wind turbine.

Nowadays most rotors are upwind i.e. the rotor faces into the wind, reducing the risk of causing localised flow instabilities that are responsible for impulsive noise. Although there are still quite a few downwind turbines (where the rotor faces away from the wind) in use, new improved design features have been incorporated aiming at reducing impulsive noise such as increasing the distance between rotor and tower.

In addition to designing upwind turbines the shape of the tower and the nacelle are aerodynamically streamlined in order to reduce any noise that is created by the wind passing the turbine.

To limit the generation of aerodynamic sounds from wind turbines the rotor's rotational speed may be restricted in order to reduce the tip speeds. Large variable speed wind turbines often rotate at slower speeds in low winds, and in increased speeds in higher winds until the limiting rotor speed is reached. This results in much quieter operation in low winds than a comparable constant speed wind turbine. Many modern wind turbines have embedded special control programs that reduce the inflow angle and rpm of the rotor depending on the time of day or year, the wind speed and the wind direction. The noise can be significantly reduced at the expense of power output.

Wind turbine blades are constantly being redesigned to make them more efficient and less noisy. The broadband tip vortex noise caused by rotating wind turbines can be tackled by giving to the blade tip an aerodynamic shape that decreases generation of vorticity. Forward sweeping into the direction of the incoming flow of the blade could result in quieter operation. Three different blade tip geometries can produce three different noise profiles.

When it comes to small wind turbines (under 30 kW) the ways to reduce noise are similar to those for large turbines. This means that they also have often variable-speed controls. The interesting fact is that small wind turbine designs may even have higher tip speeds in high winds than large wind turbines. This can result in greater sound generation than would be expected, compared to larger machines. Many modern micro wind turbines rotate over their vertical axis regulating power in high winds by turning out of the wind. This additional functionality in operation can affect the nature of the sound generation from the wind turbine during power regulation. In general domestic wind turbines apart from noise effects can generate vibration signals which are transmitted through the walls causing annoyance to people inside the house. Ways to deal with those problems are associated with increased wall insulation and with locating the turbine at a distance from the residential areas at a height over 3 m from the ground and often being kept switched off during the night.

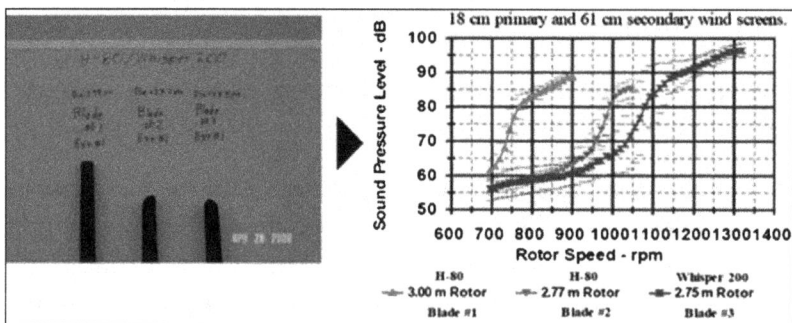

Three different blade tip geometries and the three corresponding noise levels.

Noise Standards

Currently, there are no common international noise standards or regulations for sound pressure levels from wind turbines. Every country, however, defines noise limits and regulations for human exposure depending on the time of the day.

A standard that is being internationally used is:

International Electrotechnical Commission IEC 61400-11 Standard: Wind turbine generator systems, Acoustic noise measurement techniques.

The IEC 61400-11 standard defines:

- The quality, type and calibration of instrumentation to be used for sound and wind speed measurements.

- Locations and types of measurements to be made.

- Data reduction and reporting requirements.

The standard requires measurements of broad band sound, sound levels in one-third octave bands and in narrow-bands. These measurements are all used to determine the sound power level of the wind turbine.

Construction Noise

Construction sites can have very hazardous noise levels, and they often are transient situations where different trades come to the site for short periods of time to perform their work. Depending upon the type and stage of construction, this work may be indoors, outdoors or both.

The activities and job functions at a construction site are constantly changing as the job progresses. For example, when a new building is being constructed, carpenters may build forms for the cement workers to then pour the foundation; steel workers may erect steel structures and do welding; then the building is enclosed by other workers including stucco workers, roofers and brick masons. Once the building is enclosed, carpenters, ventilation installers, electricians and plumbers begin their work, followed by drywallers, carpenters, painters and floor and ceiling men.

Each of the different trades use very different equipment to perform their jobs, and therefore, the noise created may vary. These tasks often overlap, so workers performing jobs that are relatively quiet may be exposed to noise from the other trades working around them.

Sources of Construction Noise

The main sources of noise at a construction site include construction machines (mainly machines which produce impacts, e.g. devices for breaking concrete), earth-moving machines, pile drivers, pneumatically driven devices and combustion engines. For the purposes of noise studies, these mechanisms must be considered to be point or linear noise sources depending on the level of movement at the construction site.

Some types of construction projects are not endangered by increased noise pollution at all, while others are, but only when particular conditions coincide. In the case of certain construction sites, increased impact can already be expected from the very nature and location of the project.

The latter situation mainly occurs when construction takes place in the vicinity of existing buildings or in a reflective environment. Another area which needs to be dealt with as far as noise is concerned is that of large complexes of apartment houses and family homes when the investment project is executed via a gradual construction process. In other words, construction takes place in stages which are each subject to individual approval proceedings, meaning that some buildings are already inhabited while new buildings are being created in their neighbourhood.

In order that we might work with specific values, we have to know the acoustic output value or the level of acoustic pressure at a certain distance for the evaluated source, which is information that needs to be obtained from the product sheets provided by construction machine manufacturers. Two values are often stated on a machine's technical data sheets —internal and external noise. The internal noise is the noise in the driver's cab, so it is the outside noise value that is considered for the purpose of evaluating construction site noise.

Mining and Extraction Sites

Works carried out in order to expand the productivity in mining industry have pointed out the necessity to utilise larger machinery in parallel with the improvements in technology. Increase in mechanisation has also resulted in an increase in noise levels, leading underground and open pit mines and mineral processing plants to become an enormous level of noise source. Occupational noise in underground mines has reached to unbearable levels due to the reverberant nature of the narrower spaces. Therefore, it is hard to find a relatively low-noise environment for workers. Although the equipment employed at open pits are comparatively larger in size than the ones encountered in underground, they may be said to be less significant as the noise emitted from them easily spreads hemispherically in the free sound field.

In reality, the noise occurring during the extraction works i.e. drilling-blasting, excavation; loading and transporting taking place in both open and underground pits is noteworthy when considering the labour health and job performance as the highest disease and illness rates in mining continues to be mine worker's permanent or temporary hearing loss. Additionally, it appears that noise can account for quickened pulse rate, increased blood pressure and a narrowing of the blood vessels. Workers exposed to noise sometimes complain of nervousness, sleeplessness and fatigue. Therefore, it is rather foremost to conduct a research on this matter to give suggestions to the mine management with respect to the health of workers and maximising the competence in productiveness. In comparison with the levels of noise exposure in various industries (airport, forest machinery, cement industry, foundry, textile industry, printing, metal plate workshop, ship engine room, riveting workshop), the noise levels encountered in the open cast mining industry are second only to that encountered near to jet engines at airports. Noise induced hearing loss usually occurs initially at high frequencies (3k, 4k, or 6k Hz), and then spreads to the low frequencies (0.5k, 1k, or 2k Hz).

Sources of Noise

Noise is defined as undesirable sound and it is a by-product in many industries. This is particularly true for mining. Many miners are exposed to not only loud but sustained noise levels. Most of the large excavation equipment utilised at open pits are not said to be responsible for the excessive noise levels as they are mostly equipped with noise protected operator cabs. However, excavators with lower capacity and mobile diesel-powered machines have been accepted as the primary noise sources in surface mining activities.

On the other hand, the equipment such as continuous miners, stage loaders, shearers, compressors, fans and pneumatic drilling machines may be counted as the main contributors to the excessive noise levels in underground mining. Additionally, the equipment like vibrating screens, rotating breakers and mills which are commonly in use in most of the mineral processing plants may be defined as the important sources of noise.

References

- Noise-Mapping-of-Industrial-Sources-228422573: researchgate.net, Retrieved 15 June, 2019

- Construction Site Noise and its Influence on Protected Area of the Existing Buildings, contribution to the conference proceedings Elektronický sborník konference enviBuild 2014, ISSN 1022-6680, ISBN 978-80-214-5003-5

- Noise-construction-1340: ehstoday.com, Retrieved 16 April, 2019

- Occupational-Noise-in-Mines-and-its-Control-A-Case-Study-274705857: researchgate.net, Retrieved 28 August, 2019

Chapter 4

Impact of Noise Pollution

There are various impacts of noise pollution on human health and the environment. Some of these impacts include cardiovascular disease, hypertension, sleep disturbance, cognitive impairment in children, etc. This chapter has been carefully written to provide an easy understanding of these impacts of noise pollution.

Exposure to prolonged or excessive noise has been shown to cause a range of health problems ranging from stress, poor concentration, productivity losses in the workplace, and communication difficulties and fatigue from lack of sleep, to more serious issues such as cardiovascular disease, cognitive impairment, tinnitus and hearing loss.

In 2011 the World Health Organization (WHO) released a report titled 'Burden of disease from environmental noise'. This study collated data from various large-scale epidemiological studies of environmental noise in Western Europe, collected over a 10-year period.

The studies analysed environmental noise from planes, trains and vehicles, as well as other city sources, and then looked at links to health conditions such as cardiovascular disease, sleep disturbance, tinnitus, cognitive impairment in children, and annoyance. The WHO team used the information to calculate the disability-adjusted life-years or DALYs—basically the healthy years of life—lost to 'unwanted' human-induced dissonance.

They found that at least one million healthy years of life are lost each year in Europe alone due to noise pollution (and this figure does not include noise from industrial workplaces). The authors concluded that 'there is overwhelming evidence that exposure to environmental noise has adverse effects on the health of the population' and ranked traffic noise second among environmental threats to public health (the first being air pollution). The authors also noted that while other forms of pollution are decreasing, noise pollution is increasing.

Interestingly, it may be the sounds we aren't even aware we're hearing that are affecting us the most, in particular, those we 'hear' when we're asleep. The human ear is extremely sensitive, and it never rests. So even when you sleep your ears are working, picking up and transmitting sounds that are filtered and interpreted by different parts of the brain. It's a permanently open auditory channel. So, although you may not be aware of it, background noises of traffic, aircraft or music coming from a neighbour are still being processed, and your body is reacting to them in different ways via the nerves that travel to all parts of the body and the hormones released by the brain.

Construction sites in cities add even more noise to the general traffic.

The most obvious is interrupted sleep, with its flow-on effects of tiredness, impaired memory and creativity, impaired judgement and weakened psychomotor skills. Research has shown that people living near airports or busy roads have a higher incidence of headaches, take more sleeping pills and sedatives, are more prone to minor accidents, and are more likely to seek psychiatric treatment.

But there is another, more serious outcome. Even if you don't wake up, it appears that continual noise sets off the body's acute stress response, which raises blood pressure and heart rate, potentially mobilising a state of hyperarousal. It is this response that can lead to cardiovascular disease and other health issues.

A study undertaken by Dr Orfeu Buxton, a sleep expert at Harvard University, monitored the brain activity of healthy volunteers, who were played 10-second sound clips of different types of noise as they slept. The brainwaves of volunteers were found to spike in jagged, wake-like patterns of neural activity when each clip was played. This particular study was focusing on noises heard in a hospital environment—including talking, phones ringing, doors closing, machinery, toilets flushing, and city traffic, among others—but many of the sounds tested are ones we would also hear in an urban environment.

Sound is an important and valuable part of everyday life. But when sound becomes noise, it can negatively affect our mental and physical health. The realities of modern life mean the noises created in our world are not going to suddenly fall silent. Instead, we need to recognise that noise pollution is a serious health concern worthy of our attention, and find realistic and sustainable ways to manage and reduce it—starting with banning those rubbish truck pickups in the middle of the night.

Noise-stress Relationship and Effects

The damaging effects of noise usually are regarded as limited to the structures of the ear through impairing one's ability to hear sounds such as speech and music. Often unappreciated is the fact that noise has more pervasive physiological effects.

In the course of evolution, certain fishes developed organs of hearing to orient themselves in space. In amphibians, vision provided the ability to locate prey but was not sufficient in terrestial environments to warn of other predators. Hearing accordingly developed as an organ for perceiving and responding to danger. Hearing also has played a role in sexual mating behaviors in mammals and even insects. These primitive functions exist in humans as well.

From the outset sound has evoked emotions and actions through the inner ear's direct connections to "fight or flight" neural mechanisms via the autonomic nervous system. Because of this defensive purpose, hearing also cannot be turned off, and sound registers in the brain even during sleep. Only later in primate evolution did the auditory system include higher cerebral centers permitting the appearance of spoken language.

The current usage of the terms "non-auditory" or "extra-auditory" is unfortunate. This distinction designates as non-auditory the auditory system's original, primitive influence upon wakefulness and body activity. The auditory system and physiological responses to sound are inseparably connected. Therefore, all of the effects of noise on the body mediated by the ears are "auditory" effects. More precisely, the effects of sound on the body through vibration of structures other than those of the auditory system are "non-auditory" or "extra-auditory".

Another basic consideration in understanding the functioning of the auditory system merits emphasis. The human auditory system was designed to process the frequencies and intensities relevant to survival in the sound environments of nature. The evolutionary process has not allowed humans enough time to adapt hearing to sounds generated by loud modern noise sources. This means that the auditory apparatus is not prepared to cope with commonly encountered urban and industrial noise. Consequently, we find ourselves exposed to sound environments that overload the auditory system. An analogous situation would occur in the visual system if we were forced to look at the sun and thereby damage the retina.

The fundamental relationships of hearing to emotion and action and the auditory system's vulnerability to modern sounds are not appreciated sufficiently by the public. Research also has suffered from a lack of breadth and depth in conception resulting in contradictory findings. For example, laboratory studies of healthy young people have concluded that noise has no harmful psychophysiological effects on humans. At the other extreme are reports that jet aircraft noise increases psychiatric hospital admissions.

We lack a comprehensive model to ensure that research on sound includes the critical variables that make it a significant source of human stress. Much of the research cited in this paper suffers from methodological inadequacies because noise is but one of a number of variables affecting complex human beings. Before delineating the system levels involved in bodily responses to sound, we first will outline the neuroanatomy and physiology of the central processing mechanism of sound.

Neuroanatomy of the Auditory System

An appreciation of the structural basis for physiological and behavioral responses to sound can be gained from knowledge of the neuroanatomy of the auditory system. The auditory pathways of the central nervous system consist of direct pathways from the inner ear to the auditory cortex and indirect pathways to the reticular activating system which connect to the limbic system and other parts of the brain, the autonomic nervous system and the neuroendocrine system.

The direct auditory connections consist of ascending pathways which carry impulses excited by sounds from the receptor cells in the organ of Corti to the auditory centers in the cerebral cortex. These pathways end in the temporal lobe where the sums of incoming impulses are consciously perceived and interpreted. The ascending auditory pathways travel along the auditory nerve via the cochlear nucleus, superior olivary complex, inferior colliculus, nuclei of the lateral lemniscus and geniculate body to a number of areas in the auditory cortex which in turn are connected to other cortical areas that receive inputs from the other sensory organs as well.

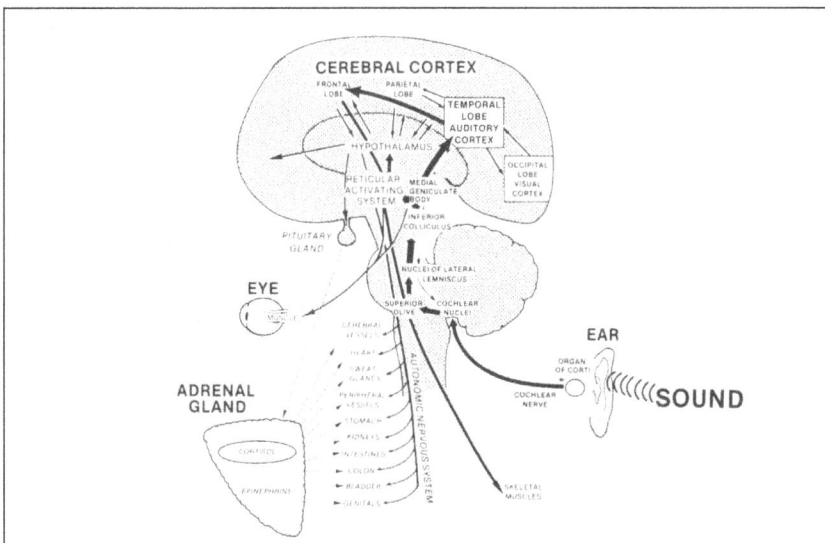

Processing of sound by the brain.

There also are descending pathways from the temporal cerebral cortex to the dorsal cochlear nucleus via the inferior colliculus and to the organ of Corti via the medial geniculate body, inferior colliculus and lateral lemniscus through the olivocochlear bundle. These descending pathways have inhibitory and, to a minor degree, excitatory influences.

In addition to these direct pathways to and from cerebral cortex, there are a variety of indirect connections from the inner ear to the brain centers that control basic physiological, emotional and behavioral responses of the body. Nerve fibers branch out from the various synaptic junctions along the direct auditory pathways to motor cell nuclei subserving reflexes within the brainstem and to the reticular activating system in the midbrain.

Impulses reaching the reticular activating system excite still other impulses that spread to higher cerebral centers that control alertness, cognition and motor performance. At the same time the reticular activating system conveys impulses to hypothalamic autonomic nervous system centers which are linked to the sympathetic-adrenal neuroendocrine system and thereby regulate the secretion of the catecholamines, adrenaline (epinephrine) and noradrenaline (norepinephrine). Impulses conveyed by the reticular activating system also are transmitted to the pituitary-adrenal neuroendocrine system which secretes corticosteroids (cortisol). The catecholamines play an important role in mobilizing immediate adaptive resources of the body, and the corticosteroids provide for more enduring adaptation to prolonged stress. Thus, the auditory apparatus is connected to the entire central nervous system and the neuroendocrine system as well.

Physiology of Sound

In conjunction with the other special senses, the auditory system serves to maintain the arousal of the brain projections to the temporal cortex and the reticular activating system via the limbic system and the hypothalamus. In this way cognitive processes and emotions interplay with sound stimuli in influencing the state of consciousness.

The cerebral cortex requires a certain level of arousal to make optimum use of incoming sensory information upon which efficient behavior and physiological functioning depend. Neither underarousal nor overarousal is conducive to effective performance of physiological functioning. Sound can improve performance on tasks which are inherently under arousing, repetitive, and monotonous. Conversely, sound can impair performance on tasks demanding concentration and complicated responses. Sound contributes to the homeostasis of the central nervous system and consequently influences the physiological homeostasis of the body through the autonomic and neuroendocrine centers of the hypothalamus.

The arousal level of the central nervous system depends upon the intensity, complexity, variability, predictability and meaning of sound stimuli. The auditory system responds most to changes in the timing of sound stimuli. Therefore, a transient increase in the firing of auditory neurons may be produced by the termination of a sound as well as by its inception. Some neurons in the auditory system respond to stimulus onset with a high rate of impulse discharge, quickly cease firing, remain silent while the stimulus is continued, and discharge a second burst of impulses when the stimulus stops. However, in a much larger number of neurons, the rate of firing declines to a lower level of activity shortly after the initial high frequency discharge and then is tonically maintained during long periods of continued stimulation. These sound stimulus-induced alterations persist after the stimulation ceases.

The direct effects of certain sounds on emotions and attitudes is illustrated by the fact that chalk scraping on a blackboard can cause chill sensations in a listener. Musical rhythm, tempo and melody can evoke moods ranging from calmness to excitation or elation. Music also can promote positive attitudes toward work. The further influence of higher cerebral cortical centers on the emotional reaction to sound stimuli is illustrated by a study of sound in hospitals in which one source of annoyance was staff conversations in the halls, not because of undue loudness but because of the discussion of patients.

Sound stimuli also influence the other sensory systems. For example, sound input overload can induce visual changes in color perception, cause nystagmus and vertigo and even act as an analgesic.

Sound stimuli play a vital role in maintaining arousal of the brain and thereby influence the basic physiological functioning of the body. Sound may influence the body after cessation of the stimulus through reverberating neural circuits within lower and higher brain centers. In this way sound can produce physiological reactions that develop a momentum of their own independent of the original stimulus.

Fundamental Auditory Responses

Orienting Response (Novelty Reflex)

The basic behavioral response to all sound stimuli is the orientation reflex, which involves ascending and descending auditory cortical pathways and is reflected by an arousal pattern in the electroencephalogram. The response orients the head and eyes toward the source of a sound in order to ready the organism to receive and respond to the sound stimulus situation. There is an associated decrease in auditory threshold and increased attention to the sound stimulus.

The orienting response occurs to sounds of low or moderate intensity and significance. The person's cognitive appraisal of the sound stimulus determines the intensity and duration of the orienting response. It extinguishes, or habituates, after varying repetitions so that the individual can accommodate to familiar and insignificant sounds with relative ease and concentrate on a preferred activity. If an appreciable amount of time passes between repetitions of a specific sound, habituation disappears and repetition of the same sound again evokes an orientation response. Habituation usually does not occur if attention to a sound is voluntarily sustained or if a sound has special significance, either positive or negative. Sounds of close to hearing threshold intensity do not easily habituate, probably because of the auditory system's difficulty in assessing their significance.

Even after behavioral habituation has occurred, sound stimuli continue to activate both cortical and subcortical areas of the brain. This is in part because excitation transmitted by the reticular activating system continues to arrive in the cerebral cortex after that transmission directly ceases. When the decision is made not to orient to a sound, descending cortical excitation actively restrains, but does not eliminate, the reticular activating system's excitation from spreading to higher areas of the brain. After the orienting response to sound habituates, there may be no change or an increase in the amplitude of the electrical responses evoked in the cerebral cortex and the medial geniculate body. The reticular activating system and the structures that it influences continue to be affected by sound even after behavioral habituation has occurred. This is not surprising because the organism's survival would be threatened by decreased alertness to danger if unattended stimuli were excluded from cognitive appraisal.

Startle Reflex

The second basic auditory response is the startle reflex which is evoked by sounds of sudden, intense, or frightening significance. The reflex has a series of components. First, the middle ear muscle reflexes via the superior olive to the tensor tympani through the fifth cranial nerve and to the stapedius muscle through the seventh cranial nerve provide a small degree of protection against sounds of extremely high intensity. The auropalpebral reflex via the superior olive and the sixth cranial nerve produces eye blinking. There also is opening of the mouth and flexion of the neck via the seventh and eleventh cranial nerves. More generally, there is flexion of most muscle groups in a "freezing" posture with rising of the shoulders, abduction of the arms, flexion of the fingers, contraction of the abdomen, and bending of the knees mediated by the ascending auditory and descending cerebral cortical motor pathways. The typical reflex is completed in less than one second.

Those components of the startle reflex that reflect cerebral cortex activity are subject to habituation or enhancement, however, those involving lower centers in the brainstem are not. The startle reflex accordingly can be decreased by anticipation, increased by background sound levels and exaggerated by emotional states such as fear.

The Defensive Response ("N" Response)

Although usually an extension of the orienting or startle responses, the defensive response merits separate consideration because it can occur independently of them and does not require sounds of high intensity. This response is produced by sounds of sufficient intensity, significance or duration to be perceived as threatening and to mobilize a "fight or flight" reaction. The response includes alerting of the cerebral cortex, emotional arousal, and preparation of the body for action.

Sounds in the range of 70 to 120 dB can produce the defensive response which appears first in the form of skeletal muscle tension that quickly reaches its peak and decays within a few seconds. Next there is a decrease in skin electrogalvanic resistance which changes more gradually than the skeletal muscle tension. Pupillary dilation occurs as well. A variety of circulatory responses are next in order: first an acceleration of pulse rate and decrease in pulse pressure, then a constriction of the finger and dilation of the chin blood vessels, followed by a slowing of pulse rate and an increase in pulse pressure. Finally in the series comes a shift to slower, deeper breathing. The defensive response also includes a reduction in salivary and gastric secretions and slowing of digestive processes.

The defensive response largely involves the sympathetic nervous system but has some parasympathetic aspects. This response is not limited to a single organ system or structural division of the nervous system. It occurs independent of emotional response on the part of the subject. It is altered by sound intensity and band width in a dose-dependent fashion. It does not completely habituate, although under laboratory experimental conditions, substantial apparent physiological habituation has been reported. It also may be elicited by low levels of sound with special significance.

Under actual working conditions, the physiological effects of the defensive response were found in sawyers exposed to bandsaw noise. Another laboratory study noted a decrease in blood eosinophile level reflecting a stress response after 25 min exposure to 85 dB level noise.

The defensive response can become the stress that leads to the General Adaptation Syndrome that will be described more fully later with its alarm, resistance, and exhaustion stages if the sound stressor is of sufficient duration, quantity, and quality. When this takes place the hypothalamicpituitary-adrenal axis is mobilized with resulting increase in adrenal cortisol and epinephrine output. During prolonged exposure to intense sounds, these endocrine effects may produce gastroduodenal ulcers and renal changes in laboratory animals.

Next we will enumerate the critical variables that determine whether or not sound stimuli become stressors that produce human stress.

Sound as a Stressor

Modern urbanization, crowding, the mass media, information technology, conditions of work and noise are overloading the human sensory environment. Of these stimuli our interest is in sound, particularly noise, although sound with meaning, such as speech, also can contribute to overloading an individual's processing capabilities. The progressive increase in noise from industrial, traffic and home sources, both machine and human generated, has reached offensive proportions in the United States.

Noise essentially is unwanted sound. As such, subjectively experienced noise is any sound that produces annoyance or communication or task performance interference.

The same sound stimulus may be perceived subjectively as noise by some and not by others. For this reason it is useful to define objectively experienced noise as sound that produces harmful bodily effects, which may or may not be subjectively perceived. This point is important because noise can be subjectively or objectively stressful, or both.

In information processing terms, noise is sound that overloads the central nervous system's prothis state can be detected by changes in the electroencephalogram. The reception of a stimulus is influenced by two kinds of cognitive state characteristics, current transient influences and enduring qualities of the individual.

The first characteristics are transient influences that are more evident and easily measured than the second type. They include level of mental arousal, from sleep through alertness to anxiety; the context of sensory stimuli arriving through the other special senses; the motor context which includes ongoing task performance, the activities of the individual; the meaning of the stimulus evoked by associations from cerebral cortical memory areas; the degree of perceived control of the stimulus, whether one is able to control the situation is helpless or expects failures and social values and attitudes toward the stimulus sources.

The level of mental arousal is influenced whether or not a sound stimulus is consciously perceived as a stressor. During the stage of early sleep, for example, sound can produce orienting and defensive responses and alter the quality of sleep without causing awakening. At the other extreme, an anxious individual can experience heightened sensitivity to a sound stimulus. For example, a study of college age males rated on an anxiety scale disclosed that for subjects rated high on anxiety, household noise levels were stressors as manifested by impaired task performance and subjective frustration.

The interaction of sound stimuli with other sensory stimuli may be significant. For example, related visual stimuli enhance the effect of sound. Clinically, sound can have an analgesic effect when certain intensities and frequencies occur in the presence of pain as is known in the practice of dentistry.

The ongoing motor activity of an individual influences cognitive state with higher levels of arousal by sound stimuli occurring while complex tasks are being performed and lower levels of arousal occurring when routine, monotonous activities are taking place.

For obvious reasons related to survival at a primitive level the meaning of sound is one of the most important factors that determine an organism's response. Threatening sounds of any kind portend potential danger; however, certain sounds acquire particular significance because of their symbolic meaning to the individual. Conversely, familiar, repetitive sounds of moderate intensity cease to attract attention. Meaning connoting potential danger then is related to unfamiliarity, rapid changes in intensity, or learned associations. For example, a study of evoked auditory potentials in the brain demonstrated that quickly changing acoustical events produce prominent cerebral excitation. The study also showed that sounds with symbolic meaning were perceived as more annoying than meaningless sounds of the same intensity and also produced larger evoked cerebral potentials.

Of particular importance is the fact that habituation does not occur to repeated novel laboratory stimuli that imply conflict or are coupled with an instruction to pay attention to that stimulus. Even covert associations with sound stimuli, such as a subject's attitude toward the experimenter, may decrease habituation. In addition, the symbolic meaning of a sound stimulus can evoke irrational responses, adding unconscious determinants of meaning.

In addition to the meaning of the sound stimulus, a sound's predictability is an important determinant of response. In one study, unpredictable noise resulted in lower tolerance for frustration and greater impairment of performance efficiency than predictable noise. Furthermore, those investigators found that an individual's ability to control the noise source, and even the belief that one could, reduced the adverse impact of unpredictable noise. They postulated that the deleterious effects of noise were a function of unpredictability and the belief that one cessing capacity because it is too great in quantity, appears too rapidly or is dissonant in meaning or pattern. Noise is a commonly used standardized stressor in laboratory testing designed to evaluate human responses to stress. In laboratory animals, for example, it is used as a stressor to produce lesions in the renal, reproductive and cardiovascular systems.

Another illustration of the use of sound is in stress studies such as the one by Cantrell, who exposed healthy young male volunteers to intermittent noise for several weeks. He found significant increases in plasma cortisol and blood cholesterol levels in addition to associated annoyance and sleep disturbance effects during prolonged exposure to bursts of 85 to 90 decibel noise.

We can use current approaches in stress research to facilitate our understanding of sound as a stressor. In stress research the environmental conditions and the intervening bodily structures and processes that determine when and in what forms stress reactions occur are taken into account.

The work of Rahe, although encompassing life change in addition to sensory stressors, is particularly useful in identifying specific variables that should be taken into account in research on the human effects of noise. Rahe developed a life stress and illness model which identifies the key steps along a pathway extending from a person's exposure to a stressor to the eventual reporting of an illness. Rahe's model utilizes the analogy of a series of optical lenses and ifiters in which stressors are depicted by a series of "light rays" of different stimulus characteristics.

The influence of a person's perceptual state in altering the experience of a stressor is represented by a "polarizing ifiter" shown in step 1. Possible sensitization, or desensitization, of a person to a stressor is indicated by changes in the "light rays" as they pass through the ifiter. The psychological defense mechanisms which appear to be capable of "diffracting away" a stressor's impact are depicted by the negative lens in step 2.

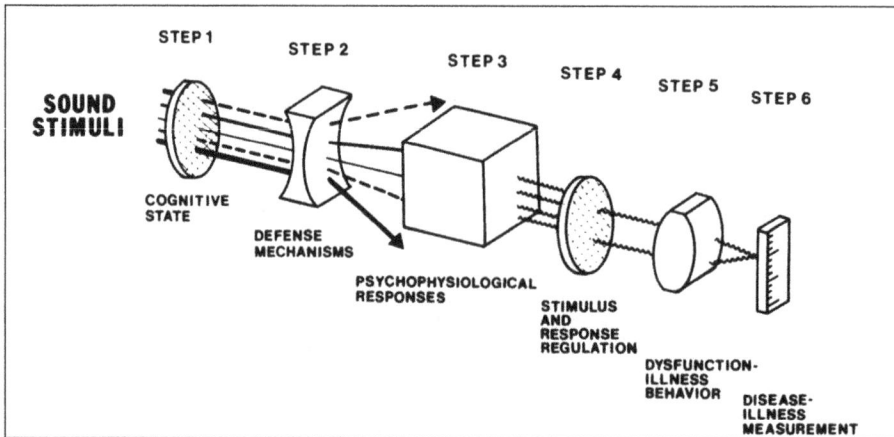

Rahe model of life stress and illness.

Stimuli not so diffracted pass on to produce a variety of physiological reactions represented by the "black box" in Step 3. The wavy lines emerging from the black box cease to represent specific stressors and begin to indicate various psychophysiological responses to perceived and "undefended" stressors. Next a "color ifiter" shown in step 4 depicts how a person may cope with or absorb certain of these physiological reactions. Prolonged psychophysiological activations, if unabsorbed, lead to organ dysfunction and eventually to psychological and bodily symptoms. Symptomatic individuals may then seek medical care. A person's "focusing" of attention on symptoms is indicated by the illness behavior "positive lens" in step 5. If these symptoms are reported to health personnel, the person receives a medical diagnosis which then can be used as a measure of illness as represented in Step 6.

The value of Rahe's model is that it incorporates pertinent system levels in conceptualizing human responses to stressors, ranging from organ system to societal levels. It permits inclusion of both subjective and objective data as well. Furthermore, the model reflects clinical realities by recognizing the social factors that influence whether or not experienced dysfunctions become labeled as manifestations of illness. For these reasons, we will use Rahe's model to elucidate key variables in the human experience of sound as a stressor.

Cognitive State

Bearing in mind the preceding discussion of the characteristics of sound experienced as noise, the first step in processing a sound stimulus is the cognitive state of the individual. Some variations in this state can be detected by changes in the electroencephalogram. The reception of a stimulus is influenced by two kinds of cognitive state characteristics, current transient influences and enduring qualities of the individual.

The first characteristics are transient influences that are more evident and easily measured than the second type. They include level of mental arousal, from sleep through alertness to anxiety; the context of sensory stimuli arriving through the other special senses; the motor context which includes ongoing task performance, the activities of the individual; the meaning of the stimulus evoked by associations from cerebral cortical memory areas; the degree of perceived control of the stimulus, whether one is able to control the situation is helpless or expects failures and social values and attitudes toward the stimulus sources.

The level of mental arousal is influenced whether or not a sound stimulus is consciously perceived as a stressor. During the stage of early sleep, for example, sound can produce orienting and defensive responses and alter the quality of sleep without causing awakening. At the other extreme, an anxious individual can experience heightened sensitivity to a sound stimulus. For example, a study of college age males rated on an anxiety scale disclosed that for subjects rated high on anxiety, household noise levels were stressors as manifested by impaired task performance and subjective frustration.

The interaction of sound stimuli with other sensory stimuli may be significant. For example, related visual stimuli enhance the effect of sound. Clinically, sound can have an analgesic effect when certain intensities and frequencies occur in the presence of pain as is known in the practice of dentistry.

The ongoing motor activity of an individual influences cognitive state with higher levels of arousal by sound stimuli occurring while complex tasks are being performed and lower levels of arousal occurring when routine, monotonous activities are taking place.

For obvious reasons related to survival at a primitive level the meaning of sound is one of the most important factors that determine an organism's response. Threatening sounds of any kind portend potential danger; however, certain sounds acquire particular significance because of their symbolic meaning to the individual. Conversely, familiar, repetitive sounds of moderate intensity cease to attract attention. Meaning connoting potential danger then is related to unfamiliarity, rapid changes in intensity, or learned associations. For example, a study of evoked auditory potentials in the brain demonstrated that quickly changing acoustical events produce prominent cerebral excitation. The study also showed that sounds with symbolic meaning were perceived as more annoying than meaningless sounds of the same intensity and also produced larger evoked cerebral potentials.

Of particular importance is the fact that habituation does not occur to repeated novel laboratory stimuli that imply conflict or are coupled with an instruction to pay attention to that stimulus. Even covert associations with sound stimuli, such as a subject's attitude toward the experimenter, may decrease habituation. In addition, the symbolic meaning of a sound stimulus can evoke irrational responses, adding unconscious determinants of meaning.

In addition to the meaning of the sound stimulus, a sound's predictability is an important determinant of response. In one study, unpredictable noise resulted in lower tolerance for frustration and greater impairment of performance efficiency than predictable noise. Furthermore, those investigators found that an individual's ability to control the noise source, and even the belief that one could, reduced the adverse impact of unpredictable noise. They postulated that the deleterious effects of noise were a function of unpredictability and the belief that one had little or no control over the noise source. A laboratory study of rhesus monkeys exposed to noise disclosed that plasma cortisol levels were significantly higher in animals with no control over the noise source than in those with control. Similar findings resulted from an experimental study of humans performing mental arithmetic problems under noise exposure.

The importance of attitude toward the noise source is illustrated by another study in which a positive or negative attitude toward all aspects of one's community consistently influenced the reporting of perceived annoyance by noise positively or negatively. In the same vein, a Swedish study disclosed that propaganda promoting the importance of the air force diminished the reported annoyance levels in a community exposed to military aircraft noise.

The second kind of variables that influence the cognitive state of an individual are enduring background characteristics in the form of individual differences in temperament and cognitive styles, organic disease processes and mental illness.

Individual variations have been demonstrated in the ways that sensory stimuli are processed. Some individuals reduce and some augment the intensity of stimuli, leading to low or high sensitivity to a particular stimulus. Sensitivity to noise also is correlated with empathic, creative, intellectually oriented personality traits, confirming Schopenhauer's comment that "noise is a torture to people of great intellect". Extraverted children may have a higher level of noise tolerance than introverted children. Moreover, individuals who express criticism tend to report annoyance by noise. Further evidence of individual differences in sensitivity to noise is reflected by the finding that some people thrive on noise which tends to synchronize their electroencephalograms while most people show electroencephalographic desynchronization. At the other extreme, it is likely that 4 to 6 % of the normal population is "noise sensitive," in the sense that they do not adapt to noise at all. For all of these types of individuals, noise has implications detrimental to their mental health.

As an illustration of other background illness characteristics, one study showed that cardiac in farction and schizophrenic patients showed greater stress responses to noise as measured by cortisol and urinary catecholamine levels than did normal subjects. Similarly, persons with cerebral vascular disease were found to be more susceptible to the detrimental effects of noise than normal subjects. Another unique group of patients harmed by sound are those susceptible to audiogenic seizures. Some are affected by sounds that produce the startle response and others by sounds such as music.

The role of psychiatric status in sensitivity to sound was suggested in a study in which normal subjects and patients with specific phobias showed habituation of physiological responses to noise while hysterical patients did not. Moreover, patients with diffuse phobias, anxiety neuroses and agitated depressions all habituated more slowly than normal. In a general sense, another survey found that psychiatric patients were more annoyed by noise than normal subjects.

Defense Mechanisms

The next clusters of variables that influence an individual's response to sound stressors are internal defense mechanisms noted in step 2 of Rahe's model. In contrast with coping techniques which are directed toward changing the stimulus environment, the defense mechanisms are devoted to maintaining homeostasis or internal equilibra, within the person.

The defense mechanisms operate automatically and unconsciously. The most primitive is the acoustic reflect which offers a small degree of protection from high intensity sounds. Another example of a physiological defense is cerebral cortical inhibition such as was demonstrated in a study of laboratory animals exposed to extreme sound which ultimately produced convulsive seizures and lethal cerebral hemorrhages. This study found that a seizure producing epileptogenic focus of excitation arising in the medulla was actively inhibited by the cerebral cortex. When this inhibitory process was exhausted, seizures occurred.

In addition to these physiological defenses, psychological defenses shield the individual from physiological arousal and also play a significant role in reducing sensitivity to sound. For example, the psychological defenses of repression and denial can minimize physiological responses as was found in a study of patients in a coronary intensive care unit.

Psychophysiological Responses

The next level, step 3 in the model, comprises psychophysiological responses to sound stressors. The psychophysiological responses can be divided into two categories. First are responses within the awareness of the individual, such as sweating, change in heart rate and muscle tension. Second are those responses which occur outside of one's direct awareness, such as changes in serum lipids, cortisol, blood pressure and blood sugar levels.

The psychophysiological responses are manifestations of the defensive response to sound, can be immediate or delayed, and occur in interactional patterns. Thus, studies of single physiological responses oversimplify the mixture of responses. An example of an immediate psychophysiological response to noise is the finding of elevated diastolic and systolic blood pressures and urinary excretion of norepinephrine metabolites in brewery workers on days in which they deliberately did not wear hearing protective devices.

Genetic and constitutional individual differences may increase the likelihood that a particular organ system will respond to stressors more than others and over time lead to disease. There also may be a critical period during infancy in which visceral learning takes place, adding conditioning of an individual's disposition to the physiological responsiveness through a specific "target" organ system. For example, in certain predisposed individuals, the target organ is the cardiovascular system, and sound stimuli produce intermittent increases in blood pressure which may eventually cause structural changes in blood vessels leading to permanent hypertension.

Stimulus and Response Regulation

The next cluster of variables is stimulus and response reduction mechanisms. These coping techniques may deal with the stressor itself or with the physiological and emotional responses to it. Using ear protective devices is an example of dealing with the stressor itself by reducing the reception of the sound stimulus.

If one becomes aware of the psychophysiological responses, particularly if they are seen as threats to health, deliberate response management techniques can be employed. For example, muscle relaxation may "absorb" the muscle tension that contributes to elevation in blood pressure.

In a broader social sense, stimulus regulation can be achieved through an individual's participation in community, industrial and consumer efforts to acoustically condition home and working environments and manufactured products.

Dysfunction: Illness Behavior

In step 5 in the model, the lack or failure of defense mechanisms and regulation techniques play important roles in producing dysfunction. The concept of sensory and information input overload is useful in understanding how the central nervous system responds when defense mechanisms and regulation techniques fail to adequately screen incoming stimuli. In this conception sensory inputs are the sound stimuli and information inputs are sounds with symbolic, message containing meaning. Overload results from an excess of the number or rate of sensory or symbolic stimuli or both.

Human experiments have shown the disorganizing and psychotogenic effects of sensory overload. Experimental exposure to intense visual and auditory sensory overload produces dramatic effects in the form of heightened and sustained arousal, mood changes, illusions, hallucinations, and body image distortions.

Sensory and informational sound overload also are commonplace in modern, urban living. Jets, air compressors, sirens, rock and roll music and road traffic are generally unpredictable and often uncontrollable sources of stimulation that contribute to making the sound of our environment inimical to mental wellbeing. Low frequency noises have effects similar to the more familiar piercing high frequency sounds.

A typical household vignette illustrates the unrecognized importance of sound sensory and informational overload in our lives. The washing machine provides a steady hum, the clothes dryer suddenly begins to vibrate; then the telephone jangles while the delivery boy rings the doorbell; a jet aircraft rumbles overhead and automobile horns are heard, a television set blares in the background; and amidst this confusion, children begin to fight, cry and scream. The overall noise level is not high by hearing damage risk criteria, but a homemaker can attest to the resulting frustration, irritability and even anger. Over time, one manages to adapt to this noise routine. However, one makes errors in balancing the check books, screams at the children for minor transgressions, is irritable with one's spouse, and generally shows symptoms of stress by the end of the day. Furthermore, when one becomes resigned to a lack of control over one's environment, the resulting "learned helplessness" itself may become a stressor and contribute to additional symptoms of depression.

Selye's classic work on stress provides a framework for understanding the body's responses to sensory and information input overload. His terms stressor and stress are comparable to stress and strain in physics. In his view stressors produce two types of changes in the body. The first is a primary nonspecific change in an organ system called the "Local Adaptation Syndrome". This local adaptation occurs repeatedly in normal living. For example, running produces stress in the musculoskeletal and cardiovascular systems. The resulting exhaustion is reversible through rest.

The second change is the "General Adaptation Syndrome" which is activated by intense and persistent stressors that produce a specific effect on the adrenal glands, thymus and stomach. The fully developed general adaptation syndrome consists of three stages: an alarm reaction, a stage of resistance and an ultimate stage of exhaustion. Extremely severe stress can lead rapidly to exhaustion and death.

Stressors, then, set in motion adaptive responses which maintain biological and psychological homeostasis. In addition to specific organ system responses, a relatively stereotyped set of neuroendocrine reactions contribute to the development of the general adaptation syndrome. Most prominent are increased secretion of the adrenal cortical hormone, cortisol, and increased activity of the sympathetic nervous system, including increased secretion of epinephrine by the adrenal medulla. The increase in sympathetic nervous system activity prepares the individual for "fight or flight". The net effect of these responses is to mobilize nutrients, such as glucose from the liver and fatty acids from fat tissue, to "arouse" the central nervous system, to provide more oxygen and nutrients to skeletal muscles, to increase contractibility of skeletal muscles and to increase coagulability of the blood. When these responses are persistent, the sustained effects of cortisol may appear in the form of gastric ulceration, inhibition of immune responses, hypertension, atherosclerosis, sterility and personality changes. Most stressors act for a limited time and produce changes corresponding to the first and second stages of Selye's syndrome. The complete general adaptation syndrome results in a specific set of physical changes: enlargement of the adrenals, shrinkage of the thymus and lymph nodes and ulceration of the stomach.

Selye's conception helps to explain that subjective experience and physiological responses to stressors can appear to have returned to normal or pre-stressor levels during the second stage of resistance. This point is essential in understanding the phenomenon of habituation which has been repeatedly observed in experimental studies of human responses to sound as a stressor. Habituation may reflect the completion of a local adaptation syndrome cycle with restoration of normal bodily functioning. On the other hand, it may reflect a stage of resistance during which the body is moving into the full general adaptation syndrome which gains a momentum of its own and exacts a physiological cost through the development of dysfunction of the various organ systems.

A stimulus appraised as threatening gradually loses its capacity to arouse an emotional response if it is reappraised on repetition as less harmful and results in adaptation. This surface adaptation may be deceptive, however, and continued exposure to the stressor may produce cumulative effects that appear after stimulation is terminated. This may be in the form of strain induced by the adaptive responses themselves. In spite of adaptation, then, stressors may cause biological and behavioral aftereffects following cessation of the stimulus.

 Laboratory studies of the physiological and behavioral reactions to noise indicate that adaptation (habituation) generally takes place in healthy subjects. There is laboratory evidence, however, which suggests that some components of physiological responses to noise do not habituate completely, although this work has been questioned. It is important to distinguish between the tension responses of an organism to stressors and stress which is a dangerous condition resulting from failure to manage tension effectively.

Another factor should be taken into account. More than lower animals, human reactions to stressors not only depend upon the direct impact of stimuli themselves but also on associated cues that signify the meaning and consequences of the stimuli. Human stress, therefore, must be defined in terms of transactions between a stimulus and an individual's reaction to the situation.

The fact that sound stimuli are processed cognitively, therefore, introduces the important conception that psychological stressors, such as the anticipation of harm, can strongly influence human responses to sound stressors. Activation of the neuroendocrine system usually depends upon the individual's recognition of a stressor as a threat. The auditory system, however, like heat or cold stressors, automatically activates the reticular activating system and thence can evoke autonomic-neuroendocrine responses.

Dysfunction resulting from sound stressors, then, may be the direct result of the sound stressor situation or may indirectly result from the activation and progress of the general adaptation syndrome. Dysfunction may be subjectively experienced as symptoms, for example, annoyance and tinnitus, or be objectively demonstrable as physical signs, for example, elevated blood pressure and hearing loss. At the same time subjective reports of interference with task performance and speech communication also can be accompanied by objective changes in cerebral responses on the electroencephalogram.

Human dysfunction can be categorized according to the influence of noise on five basic, interrelated functions that influence the well-being, or mental health, of an individual: (1) hearing, (2) sleep, (3) task performance, (4) speech communication and (5) emotional state.

Hearing

The effects of noise on hearing in the form of temporary and permanent threshold shifts and progressive deafness are well documented. There are three levels of hearing at which loss of acuity can occur: the primitive, warning, and symbolic.

The primitive level includes the familiar background sounds of one's every day environment. Loss at this level constitutes a form of sensory deprivation and may lead to a sense of isolation. Loss of hearing at the warning level contributes to a sense of insecurity and physical vulnerability in the environment. Loss at the symbolic level interferes with interpersonal communication and may result in social isolation, withdrawal and depression. Overlap between levels of hearing loss magnifies the psychological impact on the affected individual.

In addition the need to wear a hearing aid in itself may cause self-consciousness and undermine selfesteem. The emotional and psychological reactions to hearing loss accordingly are threats to an individual's mental health.

In a more general vein, the loss of hearing with aging (prebycusis) has been related in part to noise exposure in urbanized societies.

Sleep

Chronic sleep disorders detrimentally affect health and well-being. A major portion of complaints about noise arise from disturbance of rest and sleep. The Environmental Protection Agency Urban Noise survey found that 28% of the sampled population experienced sleep disturbance which also was rated as the most annoying effect of noise. Similar findings have been reported by other surveys.

The electrophysiological response to noise tends to decrease during exposure to noise during sleep; however, autonomic responses do not. This has been shown in a study in which cardiovascular responses did not habituate to traffic noise experienced by sleeping subjects. In sleep, noise evokes the same orienting response in the form of EEG arousal and changes in heart rate, GSR and finger vasoconstriction as during waking without its voluntary motor component.

The evidence also is clear that noise exposure during sleep lightens the level of sleep, especially for subjects of an anxiety-introversion personality type. Sleep disturbances have been reported in response to low frequency noise in the 20-1000 Hz range. Intermittant noise above the mean background level has a greater effect than louder, more continuous noise on vegetative functions. Age and sex in addition to sleep stage are critical factors in determining the physiological responses of noise sensitive individuals. Older subjects and women are more sensitive than younger subjects and men. Moreover, sounds with meaning tend to awaken subjects at lower intensities than those without meaning.

A study devoted to the next day effects of noise-exposed interrupted sleep showed impaired performance of tasks affected by the lack of sleep. There also is suggestive evidence that noise experienced during the waking hours may reduce the duration of sleep of susceptible persons.

All of these findings suggest that susceptible persons may be affected by noise occurring during sleep as well as during the waking state. For night workers, mothers with babies and elderly persons, day and night-time noise can be a significant problem.

Task Performance

There is little evidence that significant performance impairment on simple tasks occurs under continuous noise below 80 to 90 dB. Unpredictable or intermittent noise, however, does affect performance at even lower levels. It is well established that noise has a negative effect on work tasks that involve listening or conversing. Adverse effects occur with complex, multicomponent tasks that require prolonged vigilence or continuous performance and those in which information content is high. Under these circumstances, impairment of task performance persisting after exposure to noisy environments has been reported.

Evidence has accumulated regarding noise interference in the school performance of children. When children are involved in complex activities requiring precise movements and intense concentration, noise produces inattention and impaired task performance.

A study of the effects of noise generated by expressway traffic in homes showed higher reading achievement for children exposed to lower ambient noise levels. Another study suggests poorer task performance by young children from noisy than quiet homes.

A study compared children from schools and homes with noise levels of 87-99 dBA with those of 48-65 dBA levels. The children in the noisy environments showed increased distractibility and impaired achievement in school. Over a period of four years they became more distractible, indicating increased sensitivity to noise with the passage of time. Children exposed to high noise levels in schools also attain lower reading achievement than those with low noise levels.

One study of preschool children found that reflex motor reactions to sound and light stimuli were delayed in those exposed to moderate background levels. The children with the higher noise level required more time in task performance as well.

All of these performance effects influence an individual's attitude toward work and may constitute additional stressors in themselves, contributing to frustration and stress reactions.

Speech Communication

Interference with communication through speech not only creates personal frustration but has consequences in social interaction. Individuals react to noise levels that do not completely interfere with intelligibility. Noise accordingly may reduce efforts to converse, lead to repeated speech, and ultimately to withdrawal. The hearings impaired are particularly susceptible to these reactions. Interference with communication between teachers and children in air traffic exposed schools also has been found.

Children's speech acquisition and language development may be impaired since the repetition of speech needed to develop these skills is reduced as background noise level increases. Noise accordingly may interfere with the perception of speech by young children and affect the acquisition of language. More than adults, children depend upon the clear perception and repetition of speech sounds during early learning periods.

Emotional State

The most prominent subjective emotional response to noise is annoyance, a commonly reported but difficult to describe and quantify emotional state. Research on annoyance is hampered by the ambiguity of the term and the fact that one can be annoyed by noise itself or by its symptomatic and behavioral consequences. Arhlin relates annoyance to the direct effects, such as hearing loss, sleep, task performance and speech interference, and the indirect effects of noise, such as blood pressure elevation, headaches, fatigability, anxiety, depression and accident risk.

A standardized definition of annoyance is needed because each study tends to report annoyance in a different way. In its most specific sense annoyance is an emotion with a protective function. It motivates an individual to try to avoid or influence the sound stressor. Like discomforts such as pain, chill and warmth, annoyance serves to warn an individual of unpleasant or harmful environmental conditions. Annoyance also can be produced by interference with task performance sleep and somatic symptoms. The direct experience of annoyance by noise itself differs from indirect annoyance because of a headache.

Annoyance directly related to noise, then, is an unpleasant emotion experienced as irritability and is a form of anger or hostility related to the state of the individual in a particular social and environmental context. For example, the sound of the barking of one's own dog may be less annoying than the barking of a neighbor's dog. In another vein, some experimental subjects refused to continue in a noise study because they perceived it as unpleasant. There also is a tendency to regard sound as noise at work more than at home. Because of this relationship to the peculiarities of the context, annoyance can be expected to vary in its occurrence and reporting.

Annoyance is heightened when noise is perceived as unnecessary; when those responsible for the noise are perceived as unconcerned about the exposed population's welfare; when other aspects of the environment are disliked; when noise is believed to be harmful to health, and when noise is associated with fear.

Although defined differently from study to study, annoyance is a commonly used concept in surveys of community responses to noise. The tendency is to define annoyance of the respondents in terms of physiological responses, although a scale has been developed that excludes somatic symptoms. It can be inferred then that the more annoyed a respondent the greater the physiological reactions the person is experiencing. Direct physiological measurements would be preferable to subjective reports; however, the stage of this research is not sufficiently advanced to permit the specific measurement of noise induced stress isolated from other environmental stressors.

The evaluation of annoyance is further complicated. Not only is it an ambiguous concept, but respondents are influenced by the questions they are asked. For example, more positive responses were obtained when people were asked if aircraft noise produced specific symptoms than if asked about symptoms without suggesting a connection to aircraft noise. When annoyance was analyzed according to attitude, activity interference and symptoms, McKennel found that 65% of his sample reported feeling annoyed, 35% reported interference with activities and 5% reported symptoms.

Although methodological problems are important in assessing studies of reported annoyance due to noise, a number of surveys suggest that between 30% and 40% of urban dwellers are regularly annoyed by noise. In the flight pattern of Heathrow airport in London, 65% reported annoyance.

Noise in industrial situations also may induce what has been described as an "astheno-vegetative syndrome" in the form of increased fatigability, decreased capacity for focusing attention and slowing of motor reactions.

Disease: Illness Measurement

In step 6 of the model, whether or not people define themselves as ill depends upon individual and cultural attitudes toward assuming the patient role. These personal and social factors influence whether or not an individual minimizes or exaggerates symptoms and adopts "sick" behaviors such as missing work and seeking health care. The critical step for defining illness, then, occurs when health care is sought and a diagnosis is established. This point is the entree for measuring illness resulting from sound induced stress.

There is an inevitable gap between laboratory studies of the immediate and delayed effects of noise on health. Still, some investigators regard noise pollution in densely populated areas as a social danger comparable to that of known ingested carcinogens and air pollutants. Imposing problems, however, stand in the way of proving this thesis as illustrated by methodological criticisms of a study which found that people residing near the Los Angeles International Airport had a higher death rate from stress related diseases than a control population.

More specifically, European research on industrial noise has identified a cluster of symptoms encountered by physicians and referred to as "noise sickness". This syndrome is manifested by tinnitus, increased sensitivity to noise, fatigability, lowering of general resistance to illness, headaches, irritability, sleep interference, "heart pains," weight loss, tremors, digestive disorders and ultimately hearing loss. These symptoms are based upon the stress responses of the auditory, autonomic, cardiovascular, endocrine, and gastrointestinal systems and precede the actual loss of hearing.

Although short-term studies show that work performance can be maintained under noisy conditions, the more important consideration is the long-range effect of noise on health. The European data appear to show that complete physiological and psychological adaptation to prolonged noise exposure does not occur. The person reacts unfavorably to noise from the beginning, and adverse reactions progressively increase with the passage of time. This is most evident in work requiring complex task performance. Because of the cumulative negative effect on one's state of health, the adaptation of many individuals working in noisy environments exacts a health cost.

For convenience we will briefly summarize the relationship between noise exposure and (1) mental illness, (2) cardiovascular disorders, (3) gastrointestinal disorders, (4) neurological disorders and (5) fetal abnormalities.

Noise and Mental Illness

The role of noise in mental illness is most difficult to assess. Several studies of mental hospitals in the vicinity of Heathrow Airport in London disclosed a small but significantly higher admission rate than those in less noisy areas.

Another piece of evidence relating a form of mental illness to noise is through an indirect relationship between noises induced hearing loss and mental illness. Acquired deafness forces a change in life style toward greater social isolation and leads to an over-representation of mental illness in deaf people of all ages. One study found that 46% of a group of elderly deaf persons were paranoid and 21% had affective disorders.

On the other hand, noise-related annoyance in itself probably is not a cause of mental illness, although psychiatric patients do constitute a vulnerable group to the adverse effects of noise. Somatic and emotional symptoms associated with annoyance by noise are significant, however. The consumption of tranquilizers and sedatives has been used as an index of these symptoms resulting from noise exposure. Greater usage of these medications has been found in air and road traffic areas exposed to high levels of noise.

Noise and Cardiovascular Disorders

There is a consistent correlation between prolonged exposure to high intensity industrial noise and an increased prevalance of hypertension, as demonstrated by over 40 studies. The risk increases with advancing age and increasing years of employment for both men and women. The risk also is greater under circumstances of intermittent impulse or impact sound than continuous or relatively steady sound. There is significant confirmation under actual living and laboratory conditions of the association between high intensity sound and cardiovascular disorders in children and adults.

More specifically, prolonged noise exposure produced sustained elevations in blood pressure in a controlled experimental study of Rhesus monkeys. A carefully designed and monitored study of monkeys exposed for nine months to a continuous noise environment simulating that of urban factory workers disclosed significant, sustained elevations in blood pressure levels and alterations in diurnal blood pressure rhythms. These changes persisted after discontinuing exposure to the noise environment, suggesting a basis for long-term noise effects on the cardiovascular system in humans. Since Borg's study disclosed that lifelong exposure to high noise levels did not produce hypertension in rats, this study of primates is important because it bears a closer relationship to the human situation.

The fact that all persons exposed to noise do not show cardiovascular disorders is consistent with the likelihood that noise affects the health of susceptible individuals when combined with other stressors, such as work pressure and population density. Furthermore, environmental stressors are most likely to affect people who are unable to control them. Thus, people in institutions, with low incomes and low levels of education and children are especially likely to show adverse reactions from noise exposure.

Noise and Gastrointestinal Disorders

The data presently available are insufficient to justify judgments about the role of long term noise exposure in gastrointestinal disorders. The European literature on industrial noise, however, strongly suggests such a relationship.

Noise and Neurological Disorders

A number of investigators report neurological changes associated with long-term occupational noise exposure. The principal signs include: autonomic imbalance such as dematographism, hyperreflexia, hyperhydrosis, and hand and eyelid tremors; an altered sense of balance; decreased tactile sensitivity of the hands and feet; decreased stimulus reaction time; a decreased reactivity to visual stimuli; and obscuring of regional activity in the electroencephalogram.

An uncommon neurological disorder that is directly affected by auditory stimulation is the audiogenic seizure syndrome. In individuals with this disorder seizures are precipitated by certain sounds.

Noise and Fetal Abnormalities

The human fetus perceives and responds to environmental sound in utero as reflected in motor activity and heart rate change. In the last trimester of pregnancy, the fetus can be conditioned by external sound stimuli. Maternal anxiety related to noise can produce increased fetal activity. A possible subtle prenatal effect of maternal anxiety induced by noise exposure could be infants who are hyperactive and have dysrhythmic temperaments.

In experimental animals, maternal and fetal abnormalities have been linked to noise. At this time, however, the evidence of the adverse effect of noise on the human fetus is suggestive but not established.

The data on the health effects of noise indicate that sound exposure of more than 3 to 5 years with intensity levels of 85 dBA to 90 dBA is associated with increased health risk. Furthermore, the adverse physiological effects of noise surface before damage to hearing appear suggesting that attention to the physiological effects of noise may well enhance the prevention of noise induced hearing loss.

The effects of noise on children deserve special attention because children do not spontaneously report them, have little awareness of their significance and cannot significantly influence their environments. The evidence is that children may be particularly susceptible to noise-induced developmental and learning impairment which have long-range implications for later life.

For those who choose to question the health implications of noise, we must recognize that positive proof of cause and effect between a stimulus and human disease can never be established in the strictest experimental sense because of the multitude of intervening variables. Research in the health sciences differs fundamentally from that of the physical sciences. In even the most sophisticated epidemiological survey, a correlation remains a correlation. Species differences always exist and must be considered in even the most convincing animal study. For ethical reasons these are the only kinds of research evidence we are likely to have. We need further research to illuminate specifics but not to prove that noise is a significant threat to human health.

Environmental Noise and Sleep Disturbance

The World Health Organization (WHO) has documented seven categories of adverse health and social effects of noise pollution, whether occupational, social or environmental: hearing impairment, interference with spoken communication, cardiovascular disturbances, mental health problems, impaired cognition, negative social behaviors and sleep disturbances. The latter is considered the most deleterious non-auditory effect because of its impact on quality of life and daytime performance. Environmental noise, especially that caused by transportation means, is a growing problem in our modern cities. It is considered a major cause of exogenous sleep disturbances, after somatic problems and day tensions.

Sleep is an important modulator of hormonal release, glucose regulation and cardiovascular function. In particular slow-wave sleep, the most restorative sleep stage, is associated with decreased heart rate, blood pressure, sympathetic nervous activity and cerebral glucose utilization, compared with wakefulness. During this sleep stage, growth hormone is released while stress hormone cortisol is inhibited. Healthy sleep plays also an important role in memory consolidation. Poor sleep causes measurable changes on these systems. Experimental studies demonstrated that both sleep restriction and poor quality sleep affect glucose metabolism by reducing glucose tolerance and insulin sensitivity and that sleep restriction dysregulates appetite (lower levels of leptin and higher levels of ghrelin) as well as cortisol levels. Sleep restriction has also been shown to increase blood pressure and affect immune processes. It has been hypothesized that these perturbations cause long-term consequences on health.

Indeed there is increasing evidence that quantitative and qualitative sleep disturbances may play a role in the development of cardiometabolic disease. A number of cardiovascular risk factors and cardiovascular outcomes have been associated with disturbed sleep: coronary artery calcifications, atherogenic lipid profiles, atherosclerosis, obesity, type 2 diabetes, hypertension, cardiovascular events. Increased mortality from all causes has also been observed. During the past years, the relationship between insomnia and psychiatric disorders has come to be considered synergistic, including bi-directional causation. It has become clear that insomnia is not merely a symptom of psychiatric disorders, but contributes also to the risk of future relapse or the development of new onset mood, anxiety, and substance use disorders, as well as to the severity of psychiatric symptoms. Disturbed sleep has also been associated with increased frequency of violent acts as well as domestic violence, work and vehicle accidents, increased work absenteeism. The observed associations between poor sleep and obesity, diabetes, depression, aggressive and delinquent behaviors concern children and adolescents, too. As a result of sleep disturbances, children also suffer from impaired cognition and worsening of attention deficit hyperactivity disorder symptoms.

Nocturnal environmental noise also provokes measurable metabolic and endocrine perturbations (increased secretion of adrenaline, noradrenaline, cortisol), increased heart rate and arterial pressure, and increased motility. These biological responses to noise during sleep are most of the time unnoticed. Noise also affects sleep architecture, as well as subjective sleep quality. Nocturnal air traffic causes nocturnal awakenings at levels as low as 48 dB, and physiological reactions in the form of increased vegetative hormonal secretions, cortical arousals and body movements at even lower levels, probably around 33 dB. Nocturnal noise has been shown to fragment sleep, and as a consequence lead to a redistribution of time spent in the different sleep stages, typically increasing wake and stage 1 sleep and decreasing slow wave sleep and REM sleep, i.e. causing a shallower sleep. Basner showed that although these effects on sleep structure and continuity are relatively modest, they have a significant impact on subjective assessments on sleep quality and recuperation: Subjects experience their sleep as disturbed and with low recuperative value. Also, despite being most of the time in an unconscious state, subjects are able to distinguish between nights with low and high degrees of traffic noise exposure. Their reaction time at next day performance test is also slightly but significantly increased. These findings corroborate with previous observations that noise is indeed a widespread factor of self-reported sleep disturbances.

Apart from these measurable effects and the subjective feeling of disturbed sleep, people who struggle with nocturnal environmental noise often also suffer the next day from daytime sleepiness and tiredness, annoyance, mood changes as well as decreased well-being and cognitive performance. Associations between exposure to aircraft noise and the following health complaints and health indicators have been demonstrated: headache, poor self-rated health status, use of medication for cardiovascular diseases and use of sleep medication. Could these short-term effects be also followed by long-term adverse health outcomes? Data show that exposure to traffic noise, not specifically at night, is associated with increased incidence of diabetes, hypertension and stroke among the elderly, as well as increased incidence and mortality from coronary heart disease. But interestingly some epidemiological data support the hypothesis that exposure to noise at night time may be especially relevant in terms of negative cardiovascular outcomes, perhaps due to the fact that repeated autonomic arousals habituate to a much lesser degree to noise than cortical arousals. Indeed data show that exposure to traffic noise especially at night increases the risk for hypertension, also in children, as well as the risk for heart disease and stroke. These results confirm previous findings of studies looking at the association between subjective responses to community noise and cardiovascular outcomes that suggested that night-time noise may be more a determinant of noise-induced cardiovascular effects than daytime exposure.

Poor sleep triggers biological mechanisms contributing to the deterioration of somatic health and is clearly associated with significant psychiatric morbidity, too. Whereas exposure to traffic noise around-the-clock seems to be clearly associated with adverse health outcomes, the question of health consequences of noise exposure specifically at night still needs to be further explored, since the majority of evidence so far comes either from observational field studies that look at the immediate consequences of nocturnal noise exposure or from epidemiological studies that do not usually separate nocturnal from diurnal exposure. However, although sleep structure perturbations in the context of nocturnal noise seem less severe than in sleep pathology such as obstructive sleep apnea, they tend to be similar in their nature. It is thus reasonable to hypothesize that poor sleep may act as a mediator between nocturnal noise pollution and increased risk of cardiovascular morbidity, through impaired endocrine and metabolic functions. Also, by affecting sleep architecture, environmental noise pollution causes sleep disturbances that lead to subjective distress in the form of daytime sleepiness and tiredness, decreased well-being and cognitive performance, as well as mood changes and potentially more serious psychopathology and psychiatric morbidity, although this remains to be proven. Finally, as we have seen, by affecting biological systems in the form of a stress response causing the release of stress hormones which in turn affects factors such as blood pressure and heart rate, noise, especially at night, may also increase the risk of cardiovascular morbidity by a direct mechanism. In 2009, considering the evidence so far, a group of experts working WHO regional office for Europe officially recommended that an $L_{night, outside}$ (average level of sound pressure at night) of 40 dB (and 55 dB as an interim target) should be the target to be achieved in order to prevent nocturnal noise deleterious health consequences.

There is clear evidence that sleep disturbances are associated with health deterioration, and growing evidence that exposure to noise pollution, around-the-clock, negatively affects health, too. It has also been proven that nocturnal noise pollution significantly impairs sleep, objectively and subjectively. Whether these noise-induced sleep disturbances represent the link between environmental noise exposure and negative health outcomes still remains uncertain. However, the emerging data suggest that indeed nocturnal environmental noise may be the most worrying form of noise pollution in terms of its health consequences, possibly because of its synergistic direct and indirect (through sleep disturbances) influence on biological systems. Duration and quality of sleep should thus be regarded as risk factors or markers significantly influenced by the environment and possibly amenable to modification through both education and counseling as well as through measures of public health. One of the means that should be proposed is avoidance at all costs of sleep disruptions caused by environmental noise. Furthermore, more large scale prospective studies are needed. These studies should involve representative samples of the population including vulnerable groups like children, elderly and mentally ill subjects, have a sufficient follow-up period, assess health outcomes according to daytime versus nighttime exposure, assess hormonal and polysomnographic measures, and take into consideration potential confounders. Subgroup sleep analyses should also be performed. This would help to better understand to what extent sleep disturbances indeed mediate between exposure to environmental noise and negative health consequences.

Environmental Noise and Cardiovascular Disease

The global burden of disease has shifted within the last decades from communicable, maternal, perinatal, and nutritional causes to noncommunicable diseases, such as atherosclerosis. Although medical and scientific efforts have focused primarily on diagnosis, treatment, and prevention of traditional cardiovascular risk factors (e.g., diabetes, smoking, arterial hypertension, and hyperlipidemia), recent studies indicate that also risk factors in the physical environment may facilitate the development of cardiovascular disease (CVD). With industrialization and globalization, the importance of new environmental factors, such as noise and air pollution, is becoming increasingly evident. Within the last decade, several studies have found traffic noise (road, aircraft, and railway noise) to be associated with increased risk of cardiovascular and metabolic diseases. Already in 2011, Babisch published the statement "The question at present is no longer whether noise causes cardiovascular effects, it is rather: what is the magnitude of the effect in terms of the exposure-response relationship (slope) and the onset or possible threshold (intercept) of the increase in risk". Until recently, the precise mechanisms underlying noise-induced CVD were largely unknown, mainly because of lack of models for translational research in humans and animals. Noise annoyance and chronic stress, activation of the autonomic and endocrine system, and disturbance of sleep are proposed to ultimately lead to pathophysiologic (vascular) alterations in the intermediate or chronic timeframe contributing directly or indirectly to initiation and progression of CVD.

Adverse effects of environmental noise on the autonomic nervous system and consequences for the cardiovascular system.

According to the noise reaction model introduced by Babisch, CVD can be caused by an "indirect pathway," where lower levels of noise disturb sleep, communication, and activities, with subsequent emotional and cognitive responses and annoyance. A resulting chronic stress reaction is proposed to ultimately lead to pathophysiologic alterations in the intermediate or chronic timeframe, which may result in manifest adverse health effects. Furthermore, chronic stress may also generate cardiovascular risk factors on its own, including increased blood pressure, glucose levels, blood viscosity and blood lipids, and activation of blood coagulation, which may ultimately lead to manifest CVD. Interestingly, emotional stress induced by nighttime aircraft noise exposure has been associated with stress cardiomyopathy (Takotsubo syndrome), a phenomenon that has been linked to excessive stress hormone release. Noise-induced annoyance has been proposed to act as an important effect modifier of the relationship between noise exposure and arterial hypertension and ischemic coronary artery disease. In addition, high levels of environmental noise have been associated with mental health problems, such as depression and anxiety, conditions that are known to adversely affect cardiovascular function.

The molecular mechanisms behind the association between noise and vascular damage and CVD are not completely understood. It has been proposed that chronic stress reactions, by activation of the autonomic nervous system and increased levels of circulating cortisol, may lead to vascular (endothelial) dysfunction, mainly through induction of oxidative stress and subsequent activation of prothrombotic pathways and vascular inflammatio. In addition to endothelial dysfunction, elevated blood pressure, dyslipidemia, changes in blood glucose levels, and altered heart rate variability could contribute to CVD development or progression. Importantly, these pathophysiologic mechanisms are potentially not mutually exclusive, and may be active at different points in time following noise exposure, and vary in importance in relation to chronicity of exposure.

Adverse Cardiovascular Effects of Noise in Humans

Translational studies addressing associations between noise and vascular (endothelial) function are rare. In a recent field study, we found that simulated nocturnal aircraft noise was associated with endothelial dysfunction and decreased sleep quality. Importantly, endothelial dysfunction was markedly improved by acute administration of the antioxidant vitamin C, indicating that increased production of reactive oxygen species and depletion of antioxidant defense significantly contribute to this phenomenon. The associations between noise and endothelial function were found substantially more pronounced if the subject had been previously exposed to noise (priming effect). This indicates that the vasculature is rather sensitized than desensitized to vascular damage in response to repeated noise exposures.

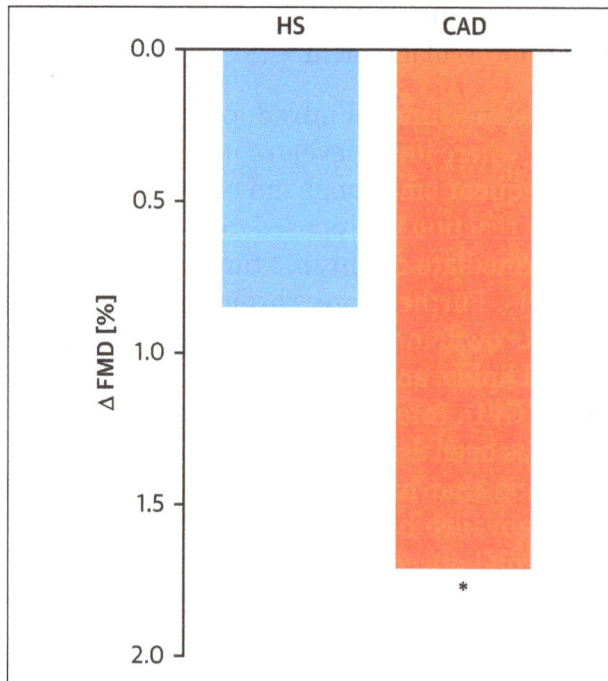

Impact of aircraft noise exposure on endothelial function of healthy subjects and patients with established coronary artery disease.

Delta in flow-mediated dilation of the brachial artery in response to 60 nighttime aircraft noise events in healthy subjects (n ¼ 70) compared with patients with established coronary artery disease (n ¼ 60). *p < 0.05 versus HS group. CAD ¼ coronary artery disease; FMD ¼ flow-mediated dilation; HS ¼ healthy subjects.

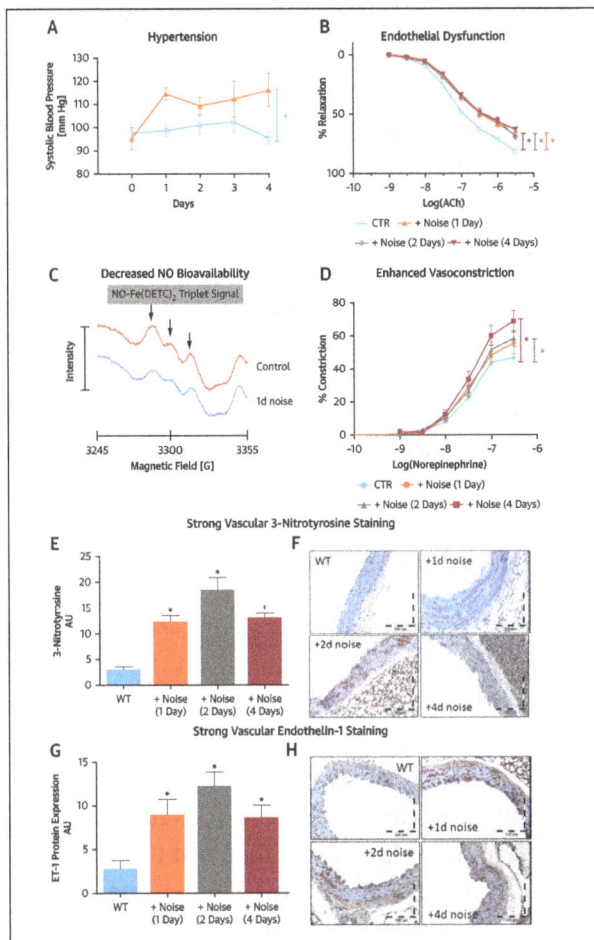

Adverse cardiovascular effects of aircraft noise exposure in mice.

Noise-induced vascular dysfunction was found paralleled by increased levels of adrenaline. Furthermore, the negative association between noise and endothelial function was more pronounced in patients with established coronary artery disease. Importantly, no correlation was observed between noise sensitivity or annoyance, suggesting that endothelial function deteriorates in response to nighttime noise, independently of whether there is an annoyance reaction or not. The study also found that simulated nighttime aircraft noise was associated with an increase in blood pressure. The Hypertension and Exposure to Noise Near Airports (HYENA) study found a significant association between nighttime aircraft noise and blood pressure. Associations between road traffic noise and CVD were found stronger among people sleeping with open windows or with bedroom facing the road. Nighttime noise may interfere with blood pressure dipping and thereby increase cardiovascular risk. Endothelial dysfunction was also observed in people working 24-h shifts and in people exposed to chronic sleep restriction, suggesting that nighttime noise–induced sleep deprivation and fragmentation may be an important branch on the mechanistic pathway between noise exposure and endothelial dysfunction and CVD.

Importantly, endothelial dysfunction has been demonstrated to have prognostic value in patients with peripheral artery disease, arterial hypertension, acute coronary syndrome, or chronic stable coronary artery disease. Thus, noise-induced endothelial dysfunction may partly explain the association between transportation noise and CVD found in various epidemiological studies.

Hypertension

Studies on chronic exposure to road traffic and railway or aircraft noise have reported a relationship with elevated blood pressure, arterial hypertension or the use of anti-hypertensive medications. These studies indicate that environmental noise may carry a considerable health burden with important medical and economic implications. A recent report from the European Environment Agency concluded that in Europe, more than 900,000 cases of hypertension are caused by environmental noise each year.

Road Traffic Noise, Blood Pressure and Hypertension

A 2012 meta-analysis of 24 cross-sectional studies on the relationship between road traffic noise and the prevalence of hypertension among adults reported an OR of 1.07 (95 % CI [1.02–1.12]) per 10 dB increase in the 16-hours daytime average road traffic noise level (L_{Aeq16h}) in the range <50 to >75 dB. A certain degree of heterogeneity among studies was detected with respect to age, gender, the way the exposure was assessed, the noise reference level used, and the duration of the exposure. For example, in the large HYENA study, road traffic noise was linked to hypertension in men but not women, and in a Dutch study road traffic noise was significantly associated with hypertension only among people aged 45–55 years. Later studies have confirmed the association between road traffic noise and prevalence of hypertension. A large Danish cohort study found a significantly higher systolic blood pressure per 10 dB increase in road traffic noise in middle-aged subjects, with stronger and significant associations in men and older subjects. No associations were found between road traffic noise and diastolic blood pressure. Similarly, a Spanish cohort study also found road traffic noise to be associated with systolic blood pressure as well as prevalent hypertension. In this study, exposure to nighttime noise was estimated outdoors as well as indoors, using information about the bedroom's orientation and indoor insulation. As expected, indoor nighttime noise levels were more consistently associated with systolic blood pressure and hypertension when compared with the outdoor levels. A major concern in studies of road traffic noise is potential confounding from air pollution. In the Spanish study, exposure to especially indoor traffic noise was associated with systolic blood pressure and hypertension independently of estimated exposure to air pollution, probably reflecting that indoor levels of road traffic noise and air pollution are less correlated than outdoor levels.

The association between transportation noise and blood pressure in children has also been investigated. A recent meta-analysis examined 13 studies comprising 8,770 children that addressed the relationship between road traffic noise and blood pressure in kindergarten and school children. The authors reported that a 5 dB rise in road traffic noise at kindergarten/school was associated with a 0.48 mmHg higher systolic blood pressure (95 % CI [−0.87–1.83]) and a 0.22 mmHg higher diastolic blood pressure (95 % CI [−0.64–1.07]). However, there was high heterogeneity in the study, and further well-designed studies are needed to further assess this association.

All these studies, both in children and adults, are of cross-sectional design, which prevents conclusions on causality and chronological order of events. Only one study has addressed the association between road traffic and railway noise and hypertension using a longitudinal design. This study found no association between long-term exposure to road traffic noise and risk of hypertension, whereas for railway noise the results indicated an association. However, this study was based on self-reported information on hypertension, which probably leads to underestimation of the actual number of subjects with hypertension, and longitudinal studies based on repeated blood pressure measurements are needed.

The effect of transportation noise on hypertension during pregnancy has also been examined. A recent Danish study found road traffic noise to be associated with hypertension among pregnant women. Based on a birth cohort including almost 73,000 pregnant women with singleton pregnancies, a 10 dB higher exposure to residential road traffic noise during the first trimester was associated with a higher risk of preeclampsia (OR 1.10; 95 % CI [1.02–1.18]) and pregnancy-induced hypertensive disorders (OR 1.08; 95 % CI [1.02–1.15]). Adjustment for air pollution lowered the estimates slightly (OR 1.08; 95 % CI [0.98–1.17] for preeclampsia and OR 1.06; 95 % CI [0.98–1.14] for all subtypes of hypertensive disorders during pregnancy). The results are comparable with results from a small Lithuanian study of approximately 3,000 women, which reported a non-significant association between road traffic noise and gestational hypertension. Although these two studies indicate that transportation noise may be a risk factor for hypertension during pregnancy, more studies addressing this end-point are warranted.

Aircraft Noise and Arterial Hypertension

An increased prevalence of arterial hypertension in the vicinity of Stockholm airport was reported in 2001. With respect to the early stages of hypertension, a time-series study in the area surrounding Frankfurt Airport showed that, even in the physiological blood pressure range, a relationship existed between aircraft noise and early-morning blood pressure. In this study, two groups exposed to nighttime outdoor aircraft noise of 50 dB(A) were followed over a period of three months. The 'western group' were exposed for 75 % of the time, and the 'eastern group' for 25 % of the time. The evaluation of ~8,000 blood pressure measurements from 53 individuals showed a statistically significant 10 mmHg higher morning systolic blood pressure and an 8 mmHg higher diastolic blood pressure for the western group compared with the less exposed eastern group.

One of the largest and most comprehensive studies on aircraft noise and hypertension is the HYENA study, based on almost 5,000 participants from six European countries. In this study, an exposure-response relationship was found, showing that for every 10 dB increase in nighttime aircraft noise (L_{night}) the prevalence of hypertension increased by 14 % (95 % CI [1.01–1.29]; P=0.031). In contrast, no effect was found for daytime aircraft noise exposure (L_{Aeq}: OR 0.93; 95 % CI [0.83–1.04]; P=0.19). Results from the HYENA study also suggest an effect of aircraft noise on the use of antihypertensive medication, but this effect did not hold for all participating study centers.

Data from the European Union-funded Road Traffic and Aircraft Noise Exposure and Children's Cognition and Health study reported an association between both daytime and nocturnal noise exposure at home and blood pressure values in children aged 9–10 years living near Schiphol (Amsterdam) or Heathrow (London). A 2009 meta-analysis of four cross-sectional studies and one cohort study on the relationship between aircraft traffic noise and the prevalence of hypertension reported an OR of 1.13 (95 % CI [1.00–1.28]; P<0.001) per 10 dB increase of the day-night weighted noise level (L_{DEN}) in the range <55 to >65 dB. This picture has been confirmed in later studies, such as a recent French study that found that 10 dB higher nighttime aircraft noise was associated with a 34 % higher prevalence of hypertension in men (95 % CI [1.00–1.97]).

Only one study has investigated the association between aircraft noise and hypertension using a longitudinal approach. This study is based on a cohort of almost 5,000 participants with repeated blood pressure measurements and living around Stockholm Arlanda airport. The authors reported that a 5 dB increase in long-term exposure to aircraft noise was associated with an 8 % increased risk for developing hypertension among men. After exclusion of the ~30 % that smoked or used snuff during or directly preceding the blood pressure measurements, this estimate increased to 21 % per 5 dB (95 % CI [1.05–1.39]). In contrast, the study indicated no association between aircraft noise and hypertension among women. The study, however, included only few subjects exposed to high levels of aircraft noise (≥60 dB), and larger prospective studies are needed in this area.

Environmental Noise and Cognitive Impairment in Children

In everyday life, cognitive tasks are often performed in the presence of task-irrelevant environmental noise. Accordingly, numerous studies on noise effects on performance have been conducted since the middle of the 20th century, showing that—depending on characteristics of sounds and tasks—noise of low to moderate intensity may in fact evoke substantial impairments in performance.

Effects of Acute Noise on Children's Performance in Auditory Tasks

Psychoacoustic studies have consistently shown that children's speech perception is more impaired than adults' by unfavorable listening conditions. The ability to recognize speech under conditions of noise or noise combined with reverberation improves until the teenage years. With stationary noise makers, signal-to-noise ratios (SNRs) have to be 5–7 dB higher for young children when compared to adults in order to achieve comparable levels of identification of speech or nonspeech signals, with adult-like performance reached at about 6 years of age. However, with maskers that vary over time, i.e., with trial-by-trial variation of the maskers' spectral composition or with fluctuating maskers such as single-talker speech, adult-like performance is usually not reached before the age of 10 years. Furthermore, children are less able than adults to make use of spectro-temporal and spatial cues for separation of signal and noise. These findings demonstrate that children are especially prone to informational masking, i.e., masking that goes beyond energetic masking predicted by filter models of the auditory periphery.

Studies identified a range of linguistic and cognitive factors to be responsible for children's difficulties with speech perception in noise: concerning the former, children are less able than adults to use stored phonological knowledge to reconstruct degraded speech input. This holds for the level of individual phonemes, as children's phoneme categories are less well specified than adults', but also for the lexical level since children's phonological word representations are more holistic and less segmented into phoneme units. Therefore the probability of successfully matching incomplete speech input with stored long-term representations is reduced. In addition, young children are less able than older children and adults to make use of contextual cues to reconstruct noise-masked words presented in sentential context. Concerning attention, children's immature auditory selective attention skills contribute to their difficulties with speech-in-noise perception. Children's susceptibility to informational masking has been attributed to deficits in focusing attention on auditory channels centered on signal frequencies, while ignoring nonsignal channels. Behavioral and ERP measures from dichotic listening paradigms provide evidence that auditory selective attention improves throughout entire childhood.

Owing to the mediating role of linguistic competence and selective attention, children with language or attention disorders are still more impaired than normally developing children by noise in speech perception tasks. A stronger noise effect is also evident for children tested in their second language when compared to native children. Studies with adults revealed that even skilled non-native listeners, whose performance in quiet is comparable to that of native listeners, are outperformed by native listeners under conditions of noise or noise combined with reverberation.

Studies reviewed so far focused on simple tasks requiring identification of isolated speech targets in noise. However, listening in everyday situations, e.g., in classrooms, goes far beyond identification of single words or syllables. Effective listening in these situations requires semantic and syntactic processing of complex oral information while developing a coherent mental model of the story meaning. Thus, the question arises how noise affects performance in complex listening tasks. Studies addressing this topic revealed noise-induced decrements in adults' memory for paired associates, sequences of unrelated words, sentences, or discourse, even with SNRs allowing perfect or near-perfect identification of the speech targets. Only a few studies in this field included children. Klatte used a listening task requiring execution of complex oral instructions and found substantial decrements due to single-talker speech and classroom noise in elementary school children. Adults were less affected. Valente reported significant impairments in discourse comprehension in 8- to 12-year-olds due to broadband noise combined with reverberation. The noise effects found in these studies could not be attributed to impaired identification. A possible explanation is that identification of degraded speech requires extra resources which are then unavailable for encoding, storage, and processing of the information. In addition, age-related improvements in attentional control may contribute to children's difficulties when performing listening tasks in the presence of noise. Children are less able than adults to ignore irrelevant sounds, and thus are more susceptible to sound-induced disruption in both auditory and non-auditory tasks.

Children need more favorable listening conditions than adults for decoding and processing of oral information. This has practical implications for the acoustical design of classrooms, since effective listening is a linchpin of school learning. The issue of classroom acoustics has thus gained much interest during the past decades. Studies simulating classroom-like conditions of noise and reverberation reported severe impairments in children's listening performance. But even though international and national standards concerning ambient noise levels and reverberation in classrooms were developed in the past decades, many classrooms still do not fit the needs of young listeners.

Effects of Acute Noise on Children's Performance in Nonauditory Tasks

Concerning tasks that do not involve auditory targets, studies with adults have consistently shown that especially short-term memory is sensitive to negative effects of noise. Immediate serial recall of visually presented verbal items is reliably impaired by task-irrelevant sounds. Impairments occur with single talker speech and non-speech sounds such as tones or instrumental music, but not with continuous broadband noise or babble noise. This so-called irrelevant sound effect (ISE) occurs reliably even with low-intensity sounds, with meaningless speech (e.g., speech in a language unknown to participants), and when sound presentation is confined to a rehearsal phase after encoding of the list items. However, the ISE magnitude is determined by inherent properties of the irrelevant sound. Recall performance is specifically impaired by sounds with a changing-state characteristic, i.e., by auditory streams which consist of distinct auditory–perceptive objects that vary consecutively. For example, irrelevant sounds consisting of different syllables or tones evoke an ISE, whereas steady state sounds, e.g., continuous broadband noise or repetitions of single syllables or tones, have a minor or no effect.

Different theories have been proposed concerning the underlying mechanisms of ISE evocation. Some of these assume that irrelevant sounds have automatic access to working memory, causing specific interference with the retention of cues to serial order or—in case of speech—with the retention of phonological codes. Other accounts attribute the ISE to the attentional burden caused by the necessity to ignore the sounds.

Several studies found the ISE in elementary school children, three of which including different age groups in order to learn about the role of attention in ISE evocation by analyzing developmental change. Elliott reported a dramatic increase in the magnitude of the ISE on serial recall of visually presented digits with decreasing age. Performance drop relative to quiet was 39% in the second-graders, as opposed to 11% in the adults. The age effect was interpreted as evidence for a dominant role of attentional control in ISE evocation. In a recent study of this group, the age effect was replicated—albeit smaller in magnitude—but other experiments in the series yielded convincing evidence against the attentional account of the ISE. Klatte used serial recall of common nouns presented pictorially and found detrimental effects due to background speech which did not differ in magnitude between first-grade children and adults. These and other findings suggest that two separate mechanisms contribute to noise-induced impairments in serial recall. On the one hand, irrelevant sounds with a changing state characteristic automatically interfere with maintenance of item or order information in short-term memory. This mechanism is the dominant source of disruption in the standard ISE paradigm, and seems to be adult-like in first-graders. On the other hand, irrelevant sounds may capture attention. The impact of attention capture depends on characteristics of the sound, and on the attentional abilities of the participants. Auditory events that are salient (e.g., of personal significance, such as one's own name), unexpected (e.g., slamming of a door), or deviant from the recent auditory context (e.g., change in voice in a speech stream) have a strong potential to capture attention. Children are more susceptible to sound-induced distraction due to limited attentional control. Accordingly, in Klatte, first-graders were also impaired by a mixture of nonverbal classroom sounds, whereas older children and adults were unaffected.

Outside the realm of research on ISE, studies addressed effects of moderate-intensity environmental noise on children's performance in academic tasks. Early studies in this field provided little support for noise-induced impairments. More recent results are inconsistent. Dockrell and Shield analyzed effects of babble and babble mixed with traffic sounds on third-graders performance in tests assessing reading, spelling, arithmetic, and attention. For all tests, overall scores were lower with babble noise when compared to quiet. Contrary to prediction, however, reading and spelling was even better in the babble plus traffic noise condition when compared to quiet and babble, and error rates in the attention test were higher in quiet when compared to both noise conditions. These results are difficult to interpret as children were not randomly assigned to noise conditions and instead were tested in their original class settings. As only two classes were assigned to each noise condition and class membership is known to affect academic performance, a-priori group differences in the dependent variables cannot be ruled out.

A number of studies investigated the effects of background speech and transportation noise on delayed memory for texts in teenagers. Participants read prose paragraphs under different noise conditions and were later tested for prose memory in silence. Recall performance was impaired by meaningful speech, but not by meaningless speech. Concerning transportation noise, results are inconsistent. Hygge found impairments due to aircraft noise during encoding. Sörqvist used a within-subjects design and found no effect of aircraft noise, but severe impairments due to meaningful speech. Hygge et al. and Hygge found impairments due to road traffic noise while Boman did not. Ljung used a direct measure of online reading comprehension and found no effect of road traffic noise and meaningful speech on 12- to 13-year olds' comprehension scores.

Thus, all except one of the studies found impairments due to meaningful speech. This is in line with studies with adults, showing that meaningful speech evokes stronger impairments than meaningless speech in school-related verbal tasks involving reading or story writing. According to the interference-by-process-account, meaningful speech automatically evokes semantic processes which compete with the semantic processes involved in the task. As transportation noise does not evoke such processes, its effect on reading found in some, but not all studies, is presumably due to a more general attention-capture process. In line with this argument, Sörqvist provided evidence that the participants' attention abilities have a stronger impact on disruption evoked by transportation noise when compared to meaningful speech. Note, however, that category membership (e.g., transportation noise vs. speech) is not sufficient to predict whether or not a sound will evoke distraction. The potential of a sound to capture attention depends on characteristics such as salience, predictability, and deviance from the recent auditory context. Thus, in addition to its specific effects on semantic processing and serial recall, speech noise containing such features is able to act as distractor. On the other hand, transportation noise lacking such features has no effect on performance.

Chronic Effects of Noise on Children's Cognitive Development

In view of the harmful effects of acute noise, the question arises whether enduring exposure to environmental noise may cause persisting deficits in children's cognitive development. Research in this field focused on indoor noise at school and aircraft noise. Concerning the former, studies yielded evidence for chronic effects on children's reading and prereading skills. Concerning aircraft noise, mixed results were reported with respect to chronic effects on children's attention and memory, but exposure to aircraft noise was consistently associated with lower reading performance. However, some of these studies are difficult to interpret due to methodological limitations. For example, cognitive abilities were usually measured in the children's regular classrooms, but acute noise levels were not always controlled. Thus, testing was done in noisy conditions for the exposed and in quiet conditions for the non-exposed children, resulting in confound of acute and chronic exposure. In addition, aircraft noise has been found to be associated with socioeconomic status (SES) which in turn is strongly related to children's reading abilities. Thus, insufficient control of SES variables in early studies may have led to an overestimation of the noise effect.

The hitherto most comprehensive study in this field, the cross-sectional RANCH (road-traffic and aircraft noise exposure and children's cognition and health) study included children (N = 2844) living in the vicinity of huge international airports in the UK, the Netherlands, and Spain. Whereas prior studies confined to comparisons of highly exposed and non- exposed children, noise exposure in the RANCH study was included as continuous variable, aiming to reveal the noise levels at which the harmful effects on children's cognition begin. With SES being controlled, the authors found no effect of aircraft noise exposure on sustained attention, working memory, and delayed recall of orally presented stories, but a linear exposure–effect relationship between aircraft noise and decreasing reading comprehension. This effect is often cited as evidence for a causal role of aircraft noise in reading impairment. What is often unreported in the secondary literature is, however, that there was another exposure–effect relationship, revealing enhanced performance in episodic memory with increasing exposure to road traffic noise. This counter-intuitive finding remains unexplained.

Concerning the underlying mechanisms of chronic noise effects, some authors proposed that enduring exposure to noise in early childhood affects the development of basic language functions which are of special importance in reading acquisition. This is a reasonable argument in view of, on the one hand, the vulnerability of children's speech perception and short-term memory for disruption due to acute noise, and on the other hand, the important role of these functions in reading acquisition. In line with this argument, electrophysiological studies revealed alterations in the cortical responses to speech sounds in individuals with a long-term exposure to occupational noise.

Environmental Noise and Tinnitus

Noise exposure is the most common cause of tinnitus. Noise induced permanent tinnitus (NIPT) can derive from occupational noise exposure, leisure noise or acoustic trauma. In general NIPT is high - pitched and tonal. The most common observed frequency of tinnitus on pitch matching is the same as the worst frequency for hearing. The sensation level of NIPT is usually low and sometimes negative. There is no correlation of significance between the discomfort caused by NIPT and audiometric findings. In occupational NIPT the interval between the start of noisy work and the appearance of tinnitus is long (many years) but with leisure noise and acoustic trauma the interval between exposure and tinnitus is frequently very short (immediate). It is a problem that the incidence of musically induced tinnitus is increasingly more common. It is also a much greater handicap for a young individual to suffer from tinnitus than from a small high tone hearing loss. Much more attention needs to be given to improve these matters. The treatment of NIPT is not different from tinnitus treatment in general.

The relationship between occupational noise and tinnitus can be assessed in different ways:

- The noise etiology in tinnitus patients.

- Noise-induced permanent tinnitus (NIPT) in relation to different types of noise and sound levels.

- NIPT in relation to NIHL and the extent of hearing loss.

The etiology for tinnitus can in most cases can only be established on the basis of the history. It is fairly unusual for tinnitus to start suddenly with a clear relationship to a specific traumatic incident; in fact it is usually slow and insidious. This of course makes any attempt to establish a clear relationship difficult and in general the etiology is based on the occupational history, any other disease with possible connection to tinnitus, audiometric results and other tests including laboratory ones. So it can be held that the etiology is mostly a "best guess". In all investigations there is a predominance of cases under the headline "uncertain etiology".

There are obviously many different types of noise in the occupational environment: pure tones, narrow-band noise, broad-band noise, impulsive noise, continuous or intermittent, low or high-pitched etc. and a large number of combinations.

There are very few investigations on the influence on hearing or tinnitus of these different physical properties of noise. The reason is the existence of a variety of noise parameters over time and the large individual variations of exposure not only due to differences in the work environment but also biological differences between individuals. Generally, it is considered that impulsive sound is more damaging to hearing than continuous noise and that pure tones are more harmful than composite sounds. Also, high-pitched sounds are probably more harmful than low-frequency sounds. Even less is known concerning the relationship between the physical properties of noise and a subsequent NIPT, but it seems reasonable to assume that factors that damage the sense of hearing also may elicit tinnitus more commonly.

A few investigators have examined specific populations exposed to occupational noise. The incidence of tinnitus in these different populations has varied much and it is probable that there are also a number of differences in the method of testing, definitions of hearing loss and tinnitus.

As can be seen the size of the populations varies much and so does the prevalence of NIPT. There is some suggestion that the incidence of NIPT is higher amongst claimants than it is in others. It seems reasonable to suggest that between 20 and 40 % of the occupationally noise exposed workers "suffer" from NIPT.

At loudness matching the sensation level of tinnitus (tinnitus loudness match - pure tone threshold) is mostly low and the most common sensation level is +5dB. Interestingly, we found that the sensation level was sometimes negative which means that the indicated tinnitus loudness was lower than the pure tone threshold. We believe that this shows that it is difficult for the patient not only to truly indicate the hearing level but also the tinnitus loudness level.

In the occupational environment there has been a pronounced change concerning noise lately, at least in the developed countries. A number of positive improvements can be listed:

- An improved awareness of the noise problems.

- An increased appreciation of preserved hearing.

- A more quiet work environment.

- Better hearing conservation programs.

- Improved technical noise abatement.

- Improved ear protectors.

- Improved use of ear protection.

- Regular and repeated industrial audiometry.

- Early identification of NIHL.

- A considerable increased public information on these matters.

As a result of this, there is reason to assume that hearing is better preserved these days and as a consequence the incidence of NIPT is decreasing. However, there are no follow-up studies available which have examined the sound levels in industry or the incidence of NIHL in noise exposed occupational populations in a regular manner over time. In spite of this we believe that noise exposure has diminished in industry with a subsequent decrease in NIHL and NIPT.

Special Case of Low-frequency Noise

Infrasound and low-frequency noise (ILFN) are airborne pressure waves that occur at frequencies ≤ 200 Hz. These may, or may not, be felt or heard by human beings. In order to clarify concepts, in this report the following definitions are used:

- Acoustic phenomena: Airborne pressure waves that may or may not be perceived by humans;

- Sound: Acoustic phenomena that can be captured and perceived by the human ear;

- Noise: Sound that is deemed undesirable;

- Vibration: Implies a solid-to-solid transmission of energy.

In the early part of the 20th century, Harvey Fletcher of the Western Electrics Laboratories of AT&T, was tasked with improving the quality of reception in the telephone. To generate the sounds in a telephone earpiece, he used an AC voltage and had some of his colleagues rate the loudness of the sound received compared to the quietest tone heard.

The company was already using a logarithmic scale to describe the power in an electrical cable and it made sense to rate the loudness of the sounds also on a logarithmic scale related to the quietest voltage that could just be heard.

Initially he called this metric a 'sensation unit' but later, to commemorate their founder Alexander Graham Bell, they renamed it the 'Bel'. A tenth of a Bel became known as the deciBel, corrupted to decibel, which has stuck with the scientific community to this day.

Fletcher-Munson Curves and the dBA Metric

To address the problem of industrial noise in the early 20th century, measurement was essential, as was a metric. At that time, researchers were critically aware that the readings on a sound level meter did not represent how loud or intense the sound was with respect to the subject's perception of hearing.

From a biomedical perspective, this concept of perception is subjective, and changes between individuals and over timescales from minutes to decades. These serious constraints notwithstanding, it was acknowledged that some average measure of loudness would have some value for medicine and public health.

Harvey continued his research with Wilden Munsen, one of his team, by varying the frequency of the electricity to give pure tones, to which it is understood 23 of his colleagues listened to different levels of loudness, again through a simple telephone earpiece. (It is assumed they all had good hearing). They were then asked to score the sounds for equal loudness to that generated by an alternating current at 1000 cycles per second.

The level of the sound of course depended on the voltage applied, which could be measured. It is important to note two significant constraints here: The sounds were 'pure' sine waves, which are not common in nature, and the headphones enclosed the ear of the subject. This is a very unnatural way to listen to a very unnatural sound.

The numerical results of this study are known as the Fletcher-Munsen Curves. The (logarithmic) units of these curves are known as 'phons' and the inverse of the 40 phon curve forms the basis of the A-frequency weighting scale used everywhere today.

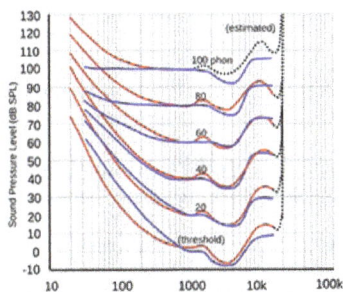

Fletcher Munson curves.

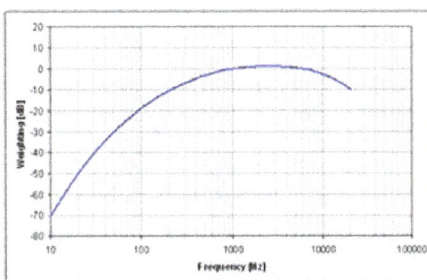

A-weighting frequency response curve.

A-frequency Weighting Scale

The minimum pressure required for humans to perceive sound at 1000 Hz is considered to be 20 micropascal, or an intensity of 10^{-12} watts per square meter. For all its shortcomings, the A-weighting has endured for decades and has become the de facto standard for environmental noise measurement. But is the A-weighting sufficient for all circumstances?

The answer is an emphatic 'No'. It relates to the perception of loudness, which heavily discounts all frequencies below 1000 Hz and ends at 20 Hz. This 20-Hz limit was a consequence of equipment limitations of the 1920s and 30s, but has remained as the lower limit of human hearing to this day. The assumption that harm from excessive noise exposure is directly related to the perception of loudness has also remained to this day.

At 10 Hz, there is a 70-dB difference between what is measured and what is, de facto, present in the environment. In other words, three-and-a-half orders of magnitude of energy are discounted at this frequency.

The implications for public health are considerable, and within this line of reasoning, any event below 20 Hz becomes of no consequence whatsoever – and more so because it is not implicated in the classical effects of excessive noise exposure: hearing loss.

There are also issues of time and frequency resolution. Acoustic phenomena are time-varying events. A 10-minute average of acoustic events can hide more than it reveals. Similarly, segmenting frequencies into octave or 1/3-octave bands for analysis can also hide much that needs to be seen.

Today, affordable and highly portable equipment can record acoustical environments, and allow for post-analysis in sub-second time increments and 1/36-octave resolution. Waveform analysis from the sound file directly can achieve an even better resolution.

References

- Exposure-response relationship of the association between aircraft noise and the risk of hypertension. Noise Health 2009; 11:161–8

- Ilfn-infrasound-low-frequency-noise-turbine-health: engineersjournal.ie, Retrieved 28 August, 2019

- Effect of chronic stress and sleep deprivation on both flow-mediated dilation in the brachial artery and the intracellular magnesium level in humans. Clin Cardiol 2004;27:223–7

- The responses of normal subjects and psychiatric patients to repetitive stimulation. In: Society Stress and Disease, L. Levi, Ed., Oxford University Press, New York, 1971, pp. 417-432

- Exposure to road traffic noise and behavioral problems in 7-year-old children: a cohort study. Environ Health Perspect 2016;124:228–34

Chapter 5

Noise Management and Mitigation

Many mitigation approaches are used for noise control and management. A few of these approaches and techniques are urban planning, noise control engineering, architectural acoustics, soundproofing, etc. This chapter closely examines these key aspects associated with noise management and mitigation to provide an extensive understanding of the subject.

The goal of noise management is to maintain low noise exposures, such that human health and well-being are protected. The specific objectives of noise management are to develop criteria for the maximum safe noise exposure levels, and to promote noise assessment and control as part of environmental health programmes. This is not always achieved. The United Nations´ Agenda 21, as well as the European Charter on Transport, Environment and Health, both supports a number of environmental management principles on which government policies, including noise management policies, can be based. These include:

- The precautionary principle. In all cases, noise should be reduced to the lowest level achievable in a particular situation. Where there is a reasonable possibility that public health will be damaged, action should be taken to protect public health without awaiting full scientific proof.

- The polluter pays principle. The full costs associated with noise pollution (including monitoring, management, lowering levels and supervision) should be met by those responsible for the source of noise.

- The prevention principle. Action should be taken where possible to reduce noise at the source. Land-use planning should be guided by an environmental health impact assessment that considers noise as well as other pollutants.

The government policy framework is the basis of noise management. Without an adequate policy framework and adequate legislation it is difficult to maintain an active or successful noise management programme. A policy framework refers to transport, energy, planning, development and environmental policies. The goals are more readily achieved if the interconnected government policies are compatible, and if issues which cross different areas of government policy are co-ordinated.

Stages in Noise Management

A legal framework is needed to provide a context for noise management. While there are many possible models, this model depicts the six stages in the process for developing and implementing policies for community noise management. For each policy stage, there are groups of 'policy players' who ideally would participate in the process.

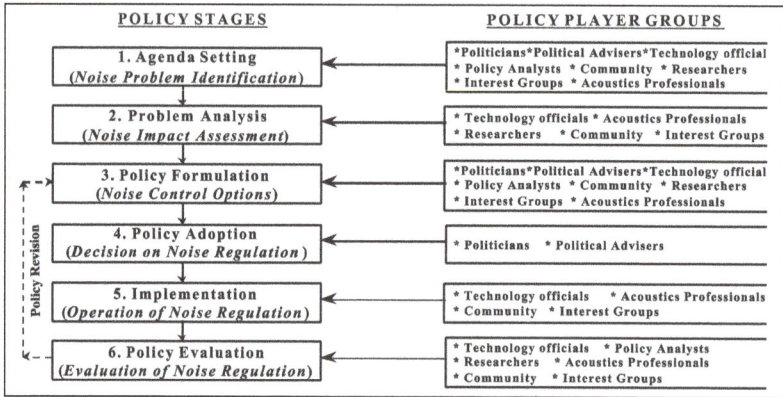

POLICY STAGES	POLICY PLAYER GROUPS
1. Agenda Setting (*Noise Problem Identification*)	*Politicians*Political Advisers*Technology official * Policy Analysts * Community * Researchers * Interest Groups * Acoustics Professionals
2. Problem Analysis (*Noise Impact Assessment*)	* Technology officials * Acoustics Professionals * Researchers * Community * Interest Groups
3. Policy Formulation (*Noise Control Options*)	*Politicians*Political Advisers*Technology official * Policy Analysts * Community * Researchers * Interest Groups * Acoustics Professionals
4. Policy Adoption (*Decision on Noise Regulation*)	* Politicians * Political Advisers
5. Implementation (*Operation of Noise Regulation*)	* Technology officials * Acoustics Professionals * Community * Interest Groups
6. Policy Evaluation (*Evaluation of Noise Regulation*)	* Technology officials * Policy Analysts * Researchers * Acoustics Professionals * Community * Interest Groups

A model of the policy process for community noise management.

When goals and policies have been developed, the next stage is the development of the strategy or plan.

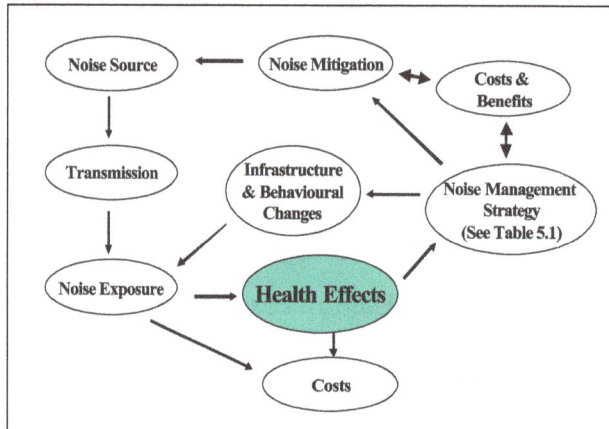

Stages involved in the development of a noise abatement strategy.

Table: Recommended Noise Management Measures.

Legal measures	Examples
Control of noise emissions	Emission standards for road and off-road vehicles; emission standards for construction equipment; emission standards for plants; national regulations, EU Directives
Control of noise transmission	Regulations on sound-obstructive measures
Noise mapping and zoning around roads, airports, industries	Initiation of monitoring and modeling programmes
Control of noise immissions	Limits for exposure levels such as national immission standards; noise monitoring and modeling; regulations for complex noise situations; regulations for recreational noise

Speed limits	Residential areas; hospitals
Minimum requirements for acoustical properties of buildings	Construction codes for sound insulation of building parts
Engineering Measures	
Emission reduction by source modification	Tyre profiles; low-noise road surfaces; changes in engine properties
New engine technology	Road vehicles; aircraft; construction machines
Transmission reduction	Enclosures around machinery; noise screens
Orientation of buildings	Design and structuring of tranquille uses; using buildings for screening purposes
Traffic management	Speed limits; guidance of traffic flow by electronic means
Passive protection	Ear plugs; ear muffs; insulation of dwellings; façade design
Implementation of land-use planning	Minimum distance between industrial, busy roads and residential areas; location of tranquillity areas; by-pass roads for heavy traffic; separating out incompatible functions
Education and information	
Raising public awareness	Informing the public on the health impacts of noise, enforcement action taken, noise levels, complaints
Monitoring and modeling of soundscapes	Publication of results
Sufficient number of noise experts	University or highschool curricula
Initiation of research and development	Funding of information generation according to scientific research needs
Initiation of behaviour changes	Speed reduction when driving; use of horns; use of loudspeakers for advertisements

The process outlined here can start with the development of noise standards or guidelines. Ideally, it should also involve the identification and mapping of noise sources and exposed communities. Meteorological conditions and noise levels would also normally be monitored. These data can be used to validate the output of models that estimate noise levels. Noise standards and model outputs may be considered in devising noise control tactics aimed at achieving the noise standards. Before being enforced, current control tactics need to be revised, and if the standards are achieved they need continued enforcement. If the standards are not achieved after a reasonable period of time, the noise control tactics may need to be revised.

National noise standards can usually be based on a consideration of international guidelines, such as these Guidelines for Community Noise, as well as national criteria documents, which consider dose-response relations for the effects of noise on human health. National standards take into account the technological, social, economic, political and other factors specific for the country.

In many cases monitoring may show that noise levels are considerably higher than established guidelines. This may be particularly true in developing countries, and the question has to be raised as to whether national standards should reflect the optimum levels needed to protect human health, when this objective is unlikely to be achieved in the short- or medium-term with available resources. In some countries noise standards are set at levels that are realistically attainable under prevailing technological, social, economic and political conditions, even though they may not be fully consistent with the levels needed to protect human health. In such cases, a staged programme of noise abatement should be implemented to achieve the optimum health protection levels over the long term. Noise standards periodically change after reviews, as conditions in a country change over time, and with improved scientific understanding of the relationship between noise pollution and the health of the population. Noise level monitoring is used to assess whether noise levels at particular locations are in compliance with the standards selected.

Mitigation Approaches at the Source and Path

The natural starting point for the reduction of environmental noise is to control the noise at source. The need to apply secondary measures, such as road design or land use, often depends on how well the control of the sources, including their number, has succeeded. However, the costs usually form a factor of prime importance. At best, noise control may be very inexpensive, if applied at the right time. But at a later stage, especially the secondary forms of noise control tend to be expensive, with the costs loaded on society or on the receiver rather than onto the user or producer of noise sources.

There are many approaches to source control. The most obvious is to control the vehicles or machines themselves through technical improvements. However, as progress is made in reducing the component sources of noise, it is also necessary to consider the number of sources and the environment where the sources operate. For instance, the road surface design can influence the noise emission, particularly at moderate and high speeds. In addition, source control can be extended to the operating conditions, e.g. driving style of road vehicles, or type of in-flight operations for aircraft.

In environmental policy, it is generally agreed that controlling noise at source is the preferred abatement method. However, additional measures which attempt to limit the spread of noise, or to give consideration to the land use near major noise sources are often needed as well. Other approaches are the improved design and insulation of buildings and non-technical (i.e. economic or social) forms of abatement.

Noise Reduction at the Source

Road Traffic: External Factors of Source Emission

Components of emission

The most important noise generating factors in motor vehicles are:

- The power train (including the engine and transmission).

- The tyre/road interaction (also termed as rolling noise).

The relative importance of the mechanisms depends on the vehicle type and driving conditions. Typically, the tyre/road noise is logarithmically related to speed: there is approximately a 12 dB level increase per doubling of speed. Power-train noise is only slightly influenced by speed. Therefore, there is a cross-over speed above which tyre/road noise dominates the overall noise.

Rolling noise has a negligible contribution to the overall noise from heavy vehicles at low speeds, but above about 30 km/h for cars and 40 km/h for heavy vehicles, rolling noise becomes a significant part of the noise. Above 50 km/h, rolling noise is the dominant noise source for cars, and above 70 km/h for heavy vehicles.

Since the range of traffic speeds in cities is spread both below and above the cross-over speed, it is obvious that both power-train and tyre/road noise must be reduced in order to obtain a better environment. In highway traffic, almost no reduction of overall noise is possible, unless tyre/road noise is reduced. Significant reductions in the noise produced by the power train are still considered feasible. However, these effects will not be fully realised for a wide range of vehicle operating conditions, unless the rolling noise components are also reduced.

Currently, and with the new generation vehicles, tyre/road noise plays a bigger role in urban traffic noise than expected before. It has been found that tyre/road noise may determine much of the overall noise even in the standard acceleration tests of new vehicles.

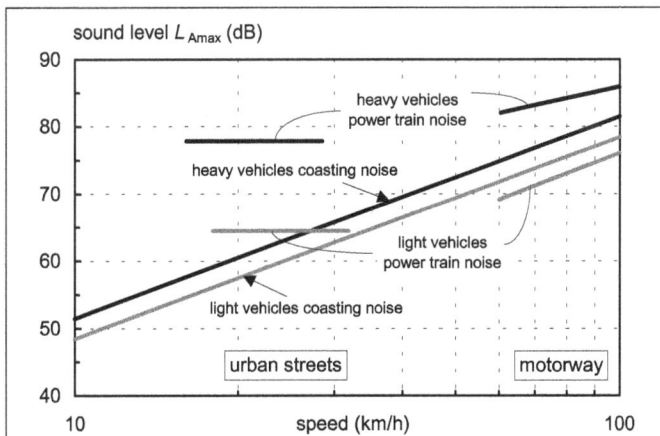

Relative significance of main noise sources in motor vehicles as a function of speed.

The most important component of rolling noise is generated by the tyres rolling over the road surface. The mechanism of noise generation is complex. At present the many factors involved are qualitatively understood. Quantitative knowledge of the interaction between the different mechanisms is, however, still incomplete.

The main factors affecting tyre noise are the speed of rotation of the tyre, the type of tread pattern and material, and the texture applied to the road surface. Earlier, it was a general believe that changes to the tyre construction and tread pattern have a smaller effect on rolling noise than changes made to the road surface material and texture. But today it is considered that there are potentially equal possibilities for controlling rolling noise by improving the road surface or by changing the tyre design.

The European Commission has invested in research with the aim to reduce the rolling noise caused by tyres. Following these studies, the Commission has made a proposal for a European Parliament and Council Directive amending Directive 92/23/EC relating to tyres for motor vehicles and their trailers. This proposal aims to reduce rolling noise by setting limit values for tyre and road noise emission and by defining their level test and measurement conditions. The final adoption of this Directive is expected during 2000.

One confusing factor in designing low-noise tyres is that today most ordinary car tyres are optimised for driving at very high speeds, up to 190–300 km/h, resulting in poor optimisation in normal conditions. A general maximum speed limit would obviously be needed on the way to quieter tyres in ordinary traffic.

Emission Limits

The noise emission limit values in use today are linked to type testing in standard conditions which emphasise the noise generated by the power train during acceleration. The limits have been rather ineffective in decreasing the emission as a whole in typical driving situations.

However, key operators of motor vehicles, such as cities and municipalities, need not adhere to common emission limits when purchasing new equipment. Stricter 'custom' limits for buses, garbage, cleaning and other working vehicles, as well as for trams and local trains may be applied as one means in noise abatement programs.

Table: The evolution of the European emission limits over time, for selected motor vehicles, LAF_{max} [dB].

Vehicle category	1972	1982	1988/90	1995/96
Passenger car	82	80	77	74
Urban bus	89	82	80	78
Heavy lorry	91	88	84	80

Low-noise Road Surfaces

Recently, open textured porous road surfaces have been developed, which offer the advantages of good skidding resistance in wet weather and good noise reduction and sound absorption characteristics. Thus these surface types provide both safety and considerable reductions of not only tyre noise but also to some extent the power train noise.

Porous road surfaces can reduce the total noise emitted by vehicles by, typically, 2–4 dB, when the surfaces are new. Up to 5 dB could be reached on high-speed roads. With suitable refinement, greater reductions up to 6–7 dB may be technically possible. This improvement seems to apply to most vehicle operating conditions, not just high speeds. In some countries (for instance the Netherlands and France), porous surfaces are already being laid as standard on a large proportion of the major road network.

However, the long-term durability is crucial, and not yet proved to be sufficient. Also, clogging can close the pores and the cleaning of the surfaces is expensive. Winter conditions are particularly rough. Especially in the Nordic Countries, icing of the pores may easily break the structure, and the use of studded winter tyres forms an additional wear factor. At present, considerable research effort is being devoted to improving the durability of this type of road surface.

Road Maintenance

Noises from the vehicle body and suspension and from movement of the load carried can be very objectionable, but these aspects are not easily dealt with. It has been suggested that some form of legislative action is needed to control this form of noise. Good road maintenance is a prime requirement for reduced load and body noise. Also, poor maintenance of roads and streets may to a large extent increase traffic noise as a whole via the increase of tyre/road noise. On the contrary, frequent re-surfacing may reduce the noise.

Road Traffic: Operational Conditions

The most important factors affecting the noise generated by the traffic stream are, besides the traffic speed, the traffic volume and the proportion of heavy commercial vehicles in the traffic flow. Another significant factor is the traffic flow, described as free flowing or interrupted as at traffic lights and junctions.

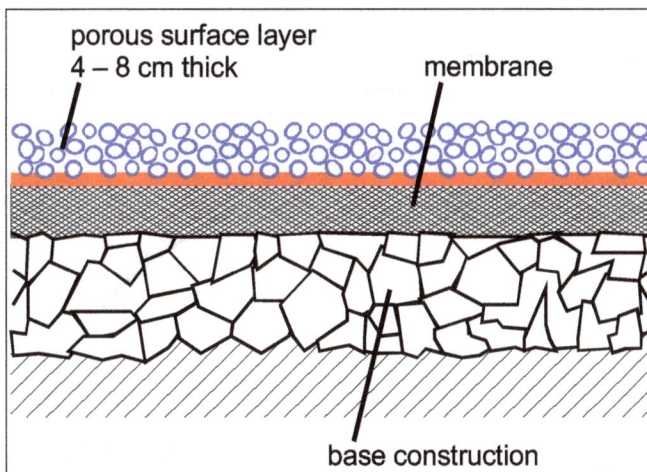

Typical structure of a porous low-noise road surface.

The foremost practical techniques to control the noise from traffic stream include re-routing, restricting access both in terms of type of vehicle, (e.g. lorry bans or preferred lorry routes), and in terms of time of day. For instance, night bans have been imposed in some areas. Control methods dealing with traffic volume and proportion of heavy vehicles form the primary steps for local authorities in reducing noise. The measures affecting speed and smooth flow are secondary steps, eventually needed to abate the noise of the remaining traffic.

Road Traffic Management

Traffic Volume and Redirection

The most obvious way to reduce traffic noise is to move the traffic away from the noise-sensitive section of the road. Concentrating urban traffic on a few main routes including by-passes, thus reducing noise in the remaining area can provide considerable benefits to many people. For example, halving the traffic flow on a residential street with light traffic may reduce noise by 3 dB. Yet the number of redirected vehicles could be quite small and easily absorbed in neighbouring roads built purposely to take higher traffic flows.

Reductions in journey times and accidents are usually additional benefits. However, closing road sections from all traffic can present problems of access, and exemptions can reduce substantially the effectiveness of the ban. On the other hand, by-passes themselves may bring along negative aspects connected to the visual landscape and access.

The effect of traffic volume controls depends not only on the proportion of traffic removed but also on the volume of traffic before and after the restrictions. Halving the traffic flow reduces noise, if other parameters do not change. However, traffic volume and speed are generally correlated and a reduction in volume is normally associated with an increase in speed. The result may be that the optimum benefits from the reduced flow are not achieved.

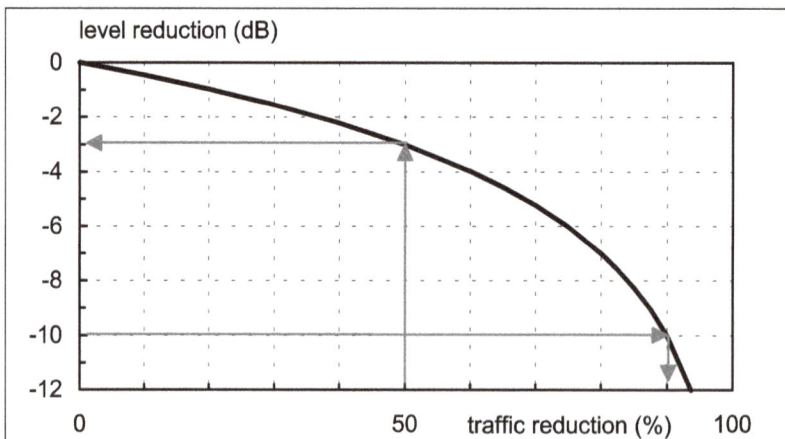

Effect of traffic volume reduction by redirection. Halving of traffic volume produces a noticeable noise level reduction of 3 dB. The subjective halving of noise (−10 dB) requires a 90 % reduction in traffic.

Removing traffic from one road produces an increase in noise on other roads. However, the fact that noise is logarithmically related to traffic volume can be used to good effect. For example, taking traffic from a lightly used road and placing it on an already heavily used road gives little increase in the noise from the latter, but the improvement on the lightly used road can be substantial.

By-passes, specifically designed to take high traffic flows to relieve residential streets from traffic, can produce considerable noise benefits. For instance, the noise reduction found in a typical case in Vienna varied between 0–4 dB. In areas where by-passes do not already exist it may be possible to use shopping streets for re-routed traffic during the night.

Restrictions in Time and Area

Restricting the number of heavy vehicles using 'sensitive' routes is one possible control measure. Generally, the noise of heavy trucks easily becomes the dominating factor in streets with low traffic speed. Bans on heavy vehicles entering a prescribed district may take the form of a total ban on all commercial vehicles above a certain capacity or weight, or restrictions at certain times, usually at night and over the weekend.

In Switzerland, heavy trucks are not allowed to run anywhere in the country during the night with the exception of buses, fire engines and trucks carrying certain perishable goods. Zürich provides an example of a combination of traffic measures to combat noise impact, involving a ban on trucks, vehicle-free zones and very quiet public transport.

Quiet zones with access limited to 'low noise vehicles' can, in principle, provide incentive both to manufacturers and operators of vehicles. It offers an attractive solution to control noise impact on a limited scale. However, potential problems may rise if standards are not harmonised between regulatory authorities. Harmonisation is similarly important internationally both for manufacturers and users.

Several Central Europe cities and regions have also introduced weekend restrictions and bans on heavy goods vehicles in order to reduce noise and gas exhaust emissions. In March 1998 the European Commission made a proposal for a Council Directive on a transparent system of harmonised rules for driving restrictions on designated roads.

In general, roundabouts produce fewer noise problems than signalized intersections. Compared to the noise of equivalent free flow traffic stream, the noise near round-abouts typically increases by about 1–2 dB.

Traffic Calming, Speed Reduction and Smooth Traffic Flow

Speed Limits

The reduction of traffic speed is in principle an effective control measure for traffic noise. On high speed roads, halving the average vehicle speed could reduce the noise by 5–6 dB. However, speed reductions cannot always be achieved easily in practice.

Speed limits are one of the most commonly used traffic restraint measure. They are generally introduced for reasons of safety. The reduction in speed caused by this restraint, if effective, will generally also lead to reduction of noise. Local speed limits are effective from a noise point of view only if they can be enforced without measures that increase the acceleration of vehicles. Another problem is that a considerable proportion of the motorists may exceed the limits. However, the limits generally tend to lower the highest speeds.

The design of the traffic speed restriction methods is important. The measures should introduce sufficient restraint on the motorist to cause speed lowering, without affecting gear changing which could result in a net increase in noise levels. The methods adopted should also ensure that traffic flows freely to encourage a nonaggressive style of driving. Speed control measures can have other positive advantages, such as reducing accidents.

The limits can also positively influence driving style and behaviour. The drivers might accelerate and decelerate less aggressively than when driving in a street with a higher speed limit. It is estimated that the noise reduction caused by driver behaviour changes may range between 2–4 dB depending upon the speed actually achieved.

Speed Monitoring

Automatic speed monitoring and even control could be one means of ensuring that designated speed limits are adhered to. Generally, there is no conflict between the goals of safety and noise abatement, when limiting the speed of road traffic is considered. The noise aspects of speeding are emphasised during the evening and night hours of low traffic, because of the eventual health effects related to sleep.

The technology for automatic speed monitoring is largely available today, including electronically enforced automatic speed control. One could argue that pressure from the environmental side might be one factor, eventually leading towards the introduction of the method within, say, one or two decades.

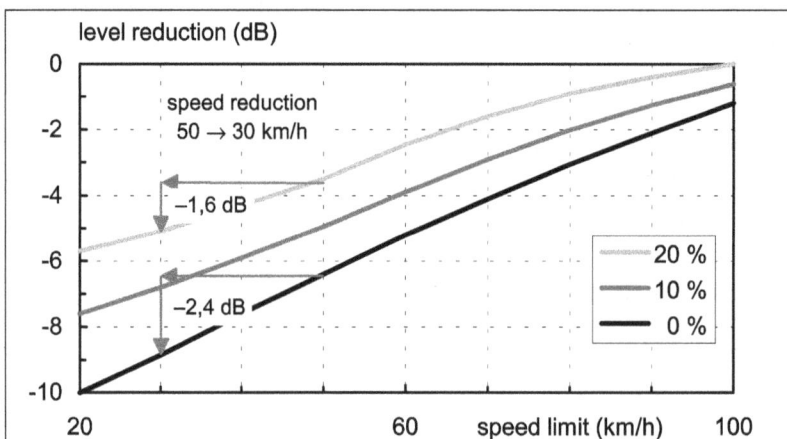

Effect of speed on noise from road traffic; parameter: proportion of heavy vehicles.

Traffic Restraints

The speed can often be reduced more efficiently by using physical restraints in addition to just speed limits. Speed can be reduced by using, for example,

- Humps laid across the road surface;

- 'Striping' of the road to give the motorist a greater awareness of speed;

- Road narrowing and road bending.

Road narrowing can be established for instance by introducing coned or widened areas for pedestrian use or dedicated bicycle lanes. Other means of narrowing can be implemented by introducing car parking bays perpendicular to the traffic stream, and road bending by varying the orientation of parking bays. These control measures are capable of substantially reducing the speeds and also the number of noisy events. Noise reductions of typically 2–3 dB can be achieved. A potential disadvantage is that too prominent restraints, such as high humps, may cause excessive braking and acceleration, and thus an increase of noise emission.

Road Junction Design

Noise can often be higher near road junctions than alongside roads with smooth traffic flow. Vehicle noise increases substantially during acceleration, particularly when the initial speed is low. At junctions, vehicles accelerate and decelerate, stop and start over. In junction design, it is important to consider how to smooth the traffic flow to minimize vehicle accelerations and thus to provide noise benefits. Also traffic management plans which aim at reducing journey times and accidents have the same objective.

For example, some noise reduction can be expected from the use of linked or demand-controlled traffic light systems. The disadvantage is that often the improved flow leads to an increase in capacity which may result in more traffic. Or an increase in traffic speed may bring increase in noise. The overall noise reductions are generally small, usually less than 2 dB.

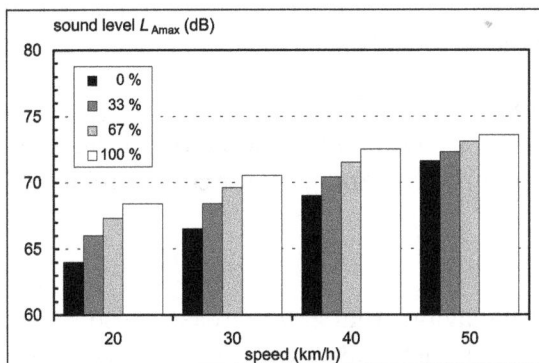

Effect of uniformity of traffic flow; parameter: unevenness.

Another measure to smooth the flow through junctions is to switch off the traffic lights at low-density junctions during the night. However, usually no systematic improvement occurs in noise since vehicle speeds are generally increased which offsets the advantages of fewer vehicles accelerating from rest.

Road Design and Alignment

The noise radiated by traffic is influenced by both the vertical and horizontal alignment of the road. Generally, the steeper the longitudinal gradient the greater the resulting noise. When designing for less steep gradients, cuttings and tunnels should reserve fair attention as an effective noise control measure and the solution to be preferred over elevated embankments and viaducts.

Driver Behaviour

The noise generated by an individual vehicle depends not only on the vehicle speed, but also the gear selected and whether the vehicle is accelerating or decelerating. These features may vary constantly as drivers attempt to cope with the traffic and road conditions. Driving styles may differ substantially; some drivers will drive less aggressively than others.

The influence of the driver on noise generation can be considerable. Therefore, the variations in driver behaviour form a potentially useful means of controlling noise. The behaviour can be affected especially when coupled with traffic restraint measures such as speed limits. Publicity campaigns and providing information can help very much both in accepting the restraints and to adopting a driving style which is effective as to noise.

In general, driving style which reduces noise also improves fuel efficiency, reduces gas exhaust emissions and improves traffic safety. Therefore, aiming at noise control supports other goals in the areas of traffic and environmental policy. Educating drivers to be more aware of the possibility of conserving fuel can thus also promote noise control. Instrumentation for fuel consumption could lead to more economical driving and less noise. Also, measures may be introduced to influence behaviour to save fuel. For example, the use of cruise control and speed limiting devices in vehicles could be useful.

Passive driving style can result in considerable fuel saving, only small increase in journey time, and substantial noise reduction. The average reduction can be approximately 5 dB for cars, 7 dB for motorcycles, and 5 dB for commercial vehicles.

Tampering

Another form of driver behaviour that can increase noise is tampering with the vehicle. This is mainly a problem associated with motorcycles. The noise increase can be as much as 20 dB. Such actions may be difficult to control via in-use enforcement. A more appropriate method is to introduce regulations which ban the sale of poor quality replacement silencers.

Rail Traffic: External Conditions at Source

A similar situation than that existing with the emission from road traffic applies generally to noise from railways as well. At low speeds the power unit contributes significantly to noise generation, but at higher speeds the interaction of wheels and rail becomes the dominant source of noise.

Rail Wheel Noise

Generating Mechanisms

The major component of train noise is caused by the interaction of the steel wheels and the steel rails, which generates sound by the vibration of wheels, rails and vehicle structure, track support and ground. The following vibration generating mechanisms may occur as a result of rail/wheel interactions:

- The impact of the wheel on a rail joint; a mechanism that is not present in the case of continuously welded rail (reduction 3–5 dB).

- The impact of wheel flanges against the rail.

- The motions caused by track and wheel irregularities: corrugations in the rail and flats on the wheel as well as smaller scale roughness in both rail and wheel. Can raise the noise level by 10–20 dB.

- Sliding contact of wheel flanges resulting in flange squeal (control measure is to avoid tight radii of curvature). 5. Vibration of the supporting structure.

Control Methods

Methods of controlling rail/wheel noise generally follow two directions. Besides actions on the train itself, it may be possible to reduce or control the roughness of both wheels and rails and to reduce the formation of wheel flats and rail corrugations. For example, it has been found that both wheel flats and rail corrugations may be substantially reduced by employing disc brakes rather than the more conventional wheel tread braking. Also, rail corrugations can be controlled by routine grinding of the rails.

Other methods of control include the use of bogie skirts, railside noise barriers, reducing the number of wheels and the employment of rail isolation techniques. These include the use of resilient rail fasteners to aid damping in the rail and to decouple the rail from the support structure, and the use of ballast mats on bridge decks to limit vibration coupling to the bridge structure. In addition, enhancing the ground absorption between and beside the rails may further reduce the noise by a few decibels.

Rail Maintenance

The conditions of the surface of the rail and the tread of the wheel have a significant effect on noise from a train. Defects in the wheel tread such as flats (due to wheels sliding during braking), loss of portions of the wheel tread due to thermal or mechanical fatigue, various rail running surface defects, and rail joints are all major causes of train noise.

Air Traffic: Operational Controls

Over the last decades the problem of noise from aircraft has received substantial attention and some major successes in controlling it has been achieved. The methods dealing with the abatement of the noise emitted can again be divided into two:

- Development of quieter power units;

- Regulation and control of aircraft operation in the vicinity of airports.

Also the number of sources (volume of air traffic) affect the overall noise reduction at the receiver. For instance, the noise emission from aircraft has decreased during the past 20 years more than 10 dB, but the increase in traffic volume results in roughly no reduction as a net effect.

The noise emitted by the aircraft themselves have been subject to continuously tightening emission limits for a number of years. The tightest limits were declared some 25 years ago, and no further lowering of the emission limits have been agreed upon.

The European Union has been concerned with the lack of further progress on a new noise stringency standard. Consequently, The EU Council has adopted Regulation 925/99 on the registration and operation of modified and recertificated aircraft (i.e. equipped with the so called 'hush-kits').

In-flight Operations

The noise of an aircraft is closely related to the way it is operated. For instance, the noise can vary considerably depending on the climb-out or landing approach procedures used. In addition, ground operations can influence airport noise impact.

Takeoff/Landing Restrictions

Aircraft are at their loudest during takeoff when full power is used. If residential or other sensitive areas are situated close to the airport, the aircraft have not climbed to a sufficient height when crossing them and noise reduction measures are needed. For example, engine power is reduced when a safe height is reached and the climb is continued more gradually.

Under an approach route to an airport the noise may be comparable although the engine power is lower than at takeoff. The rate of descent must be low to keep it within safe limits. Typical angle of descent for the last phase of approach is 3° from an initial height of 600 m.

Both landing and take-off restrictions are able to reduce the size of an airport noise contour. Many air carriers have adopted a noise abatement procedure as a policy wherever they operate.

At many airports the reverse thrust is used when landing to reduce aircraft speed. However, this produces a loud noise event for a short period, and at airports with sensitive areas nearby can give rise to noise problems. Some airports apply thrust reversal controls as part of their noise abatement plan and several airports do not permit reverse thrust procedures due to noise.

Noise Monitoring

At most major airports climb-out is controlled using noise monitoring stations under the departure routes. They indicate whether the prescribed climb profile and engine power settings are followed.

Noise Abatement Flight Tracks

Flight-track procedures with unique flight tracks prescribed for departures or arrivals are commonly used for avoiding flight over noise-sensitive areas. The flight track is the projection onto the ground of the flight path of aircraft. For noise abatement purposes, the flight tracks can be useful for both approach and departure in positioning the aircraft relative to ground or land uses.

Many airports assign headings that place aircraft over nonpopulated land including water; or agricultural and wilderness areas. However, for airports with a dense surrounding population, this kind of flight track method may not be applicable. Then a rotation of operating runways may be used, with flight tracks distributed in a more or less equal pattern, in order to attempt to spread the noise geographically more evenly to surrounding communities.

Time of Day Restriction (Curfews)

Air controls of this type generally apply to the time aircraft are permitted to operate, typically limiting the hours in which an airport may permit flight operations. Many major airports have some form of restriction during the night. For example, no jet operations are allowed at night in certain specified areas (Washington), or the number of take-offs during the night is substantially restricted. In a region with more than one jet airport, a night curfew can be a feasible proposition. Switzerland imposes a night-time operational curfew for all traffic.

A partial curfew is also common where the airport permits certain operations at night, based upon the type or class of aircraft. For example, scheduled departures of noisier aircraft may be prohibited. Some airports place a restriction on the total number of operations over a selected long time period, for instance summer. Restrictions in many countries are not limited to only civil aircraft. In rare instances complete curfews are in effect restricting any aircraft, usually at night.

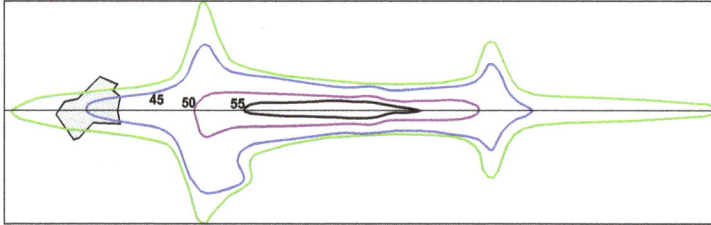

Example of the noise zones of an airport as a result of noise control actions.

In the European Union, discussions have taken place on the possibility of banning the whole of air traffic during the night (hours 23–06). One disadvantage from this action is a difficulty of scheduling flights with longer distances, perhaps involving multiple time zones.

Perimeter Rule

This rule is used to limit the stage length of flights departing from and arriving at the airport. It is sometimes applied when there are nearby airports that can operate without such a restriction. This can influence noise in several different ways:

- It can impact the capacity of an airport. In general, the fewer operations limit the overall noise.

- With restricted lengths the maximum take-off weight, heavily influenced by fuel, is less. This permits more lift and can reduce the ground noise contour of the aircraft.

- The aircraft type needed for a reduced stage length may be quieter than aircraft used for longer flights.

Ground Operations and Traffic Management

Aircraft ground operations can also cause noise problems close to airports. The sources of ground operation noise include engine testing and run-up prior to taxiing, and standing aircraft noise on apron and terminal stands. Methods of controlling the noise from these operations include, for run-up noise, reorientating or relocating the aircraft away from noise-sensitive areas or the use of suppressors and barriers. Other ground operations are controlled using space to separate noisy operations, such as start of roll, from sensitive areas, and buildings and screens to shield the noise.

The management of the airport can have a significant influence on noise control. Airport design and its operational runways, run-up areas, buildings and noise barriers can influence community impact. Administrative controls and remedies, such as operationally based charges for noise, are increasingly common.

The foremost ground-based control methods include:

- Slots, limiting the number of operations within a specified time period. This can involve restrictions on how many air carrier movements are allowed during a 24-h period.

- Capacity generally refers to the number of flights or passengers permitted over a defined period, e.g. a year. A major reason is to limit the noise.

- Preferential runways, the most common operational technique in use.

- Displaced threshold. A point along the runway is used as a landing threshold for arriving aircraft or a take-off point for departing aircraft. If displaced, the approach or departure takes place higher above the ground. Increasing the altitude reduces the noise exposure contour, thereby affecting fewer people.

- Ground run-up. Static run-up tests of engines, associated with maintenance and repair, can generate noise impact, depending upon location, time of day, aircraft type, and facility. Run-up noise control is usually achieved applying the measures of noise emission limit, time of dayrestrictions, location, and site design or abatement equipment.

- Aircraft towing. The towing of aircraft for noise control purposes is not common, but the technique may reappear, depending upon safety, energy, and noise cost-benefits in the future.

- Noise based charges, the idea being that the aircraft operators should pay a fee proportional to the generated noise. The operators of noisier aircraft are financially penalized while the operators of quieter aircraft are rewarded by reduced landing charges.

Training Restrictions

Aircraft training activities are an important subject for regulation in many airports, and training restrictions are a common type of operational noise control method.

Industry: Control at Source

Industrial noise is generated by a wide and mixed collection of various types and forms of noise sources. Therefore, only a very general outline of their abatement measures can be drawn.

One typical class of sources are rotating machinery: fans, pumps, compressors as well as gas turbines, diesel and electric motors, and gears connected to these. Another is formed by moving or flowing gas, liquid or solid particles, in ducts, pipes, transmission lines, through valves, at openings into open air etc. A third class may be working machinery, often impulsive as to noise.

Specific examples of strong concentrated sources are stone crushing plants, stone quarries, pithead installations and asphalt stations. Power stations are also often considerable noise sources. Wide area-type outdoor sources may be found in petrochemical complexes and saw mills. A general 'factory' may radiate noise from machinery located outside or on the roof of the building, or through walls and openings. Moving machinery, fork-lift trucks, excavators or cranes may operate outdoors, etc.

A rough list of typical control measures at source may include the following:

- Silencers, attenuators or mufflers, in connection with rotating machinery and ducts/pipes leading to/from these;

- Screens, enclosures, cladding or even dedicated separate huts or buildings;

- Improved sound insulation of buildings, for walls, windows, doors, openings, ventilation etc;

- Alternate, inherently less noisy solutions or devices in processes or production procedures (such as lower velocity, screwing or cutting instead of striking etc).

Permission policy adopted by the (usually local) authorities is a general means for abating or controlling industrial noise.

Engineering Controls

Engineering controls modify the equipment or the work area to make it quieter. Examples of engineering controls are: substituting existing equipment with quieter equipment; retro-fitting existing equipment with damping materials, mufflers, or enclosures; erecting barriers; and maintenance.

Administrative Controls

These are management decisions on work activities, work rotation and work load to reduce workers' exposure to high noise levels. Typical management decisions that reduce worker exposures to noise are: moving workers away from the noise source; restricting access to areas; rotating workers performing noisy tasks; and shutting down noisy equipment when not needed.

Personal Protective Equipment

Earplugs are the typical PPE given to workers to reduce their exposure to noise.

Earplugs are the control of last resort and should only be provided when other means of noise controls are infeasible. As a general rule, workers should be using earplugs whenever they are exposed to noise levels of 85 dB (A) or when they have to shout in order to communicate.

Construction Sites can be Quieter

Although many in the construction industry believe that construction sites are inherently noisy, there are many ways in which they can be made quieter.

- Sometimes a quieter process can be used. For example: Pile driving is very loud. Boring is a much quieter way to do the same work.

- New equipment is generally much quieter than old equipment. Some equipment manufacturers have gone to great lengths to make their equipment quieter. Ask equipment manufactures about the noise levels of their equipment and consider these levels when making your purchase. For example, noise-reducing saw blades can cut noise levels in half when cutting masonry blocks.

- Old equipment can be made quieter by simple modifications, such as adding new mufflers or sound absorbing materials.

- Old equipment is also much quieter when it is well maintained. Simple maintenance can reduce noise levels by as much as 50%.

- Noisy equipment can be sited as far away as possible from workers and residents. Noise levels drop quickly with distance from the source.

- Temporary barriers/enclosures (e.g. plywood with sound absorbing materials) can be built around noisy equipment. These barriers can significantly reduce noise levels and are relatively inexpensive.

Urban Planning for Noise Management

Classical urban noise management mainly aims at reducing the level of unwanted noise - of mechanical or electronic origin - in the city by source mitigation and by obstructing the propagation path. Remediation after the problem arises or during planning in a more ideal situation using technical measures, is by far the most common approach to noise control. No doubt this approach is a very good one provided that it leads to significant noise reduction. However, this is certainly not the case in many of our modern cities and therefore more creative approaches are necessary. This chapter reports on some recent developments in this area and puts the urban soundscape and its design in a slightly different perspective.

For the reader less familiar with traffic noise management we summarize the most relevant technological possibilities for urban noise control before tackling the problem in detail. Road traffic noise is caused by the drive train at the one hand and by the wheel-road interaction (rolling noise) at the other. As driving speed increases and driving is more regular, the latter contribution starts to dominate. Drive train noise is constrained by European noise emission regulations which have progressively become more restrictive over the years. Manufacturers have modified many technical details of cars to comply with these regulations, but they did not necessarily do much more than just that. Cars and buses powered by alternative fuel, in particular electric engines, can be significantly quieter. Although the European noise regulation cannot be circumvented by member states, they have opportunities to stimulate citizens and enterprises (including public transport companies) to buy more environmentally friendly vehicles or to introduce noise testing for vehicles in use. Rolling noise involves tires and road surface. Tire requirements, including noise, are also a European matter. A good choice of road surfaces can significantly reduce rolling noise, but road surface maintenance is at least as important, in particular in the urban context. In (VMM, 2007a) it is shown that there is some potential for reducing rolling noise in Flanders both through modifying road surfaces and through modifying tires, but the number of dBs to be gained is small.

The picture for railway noise is not all that different. Many rail vehicles are electrically powered and thus engine noise is often limited. Rolling of metal wheels on metal rails produces a level of sound that depends on the smoothness of both the rail and the wheel. The opportunities for noise reduction at the source are probably more significant than for road traffic, in particular when considering freight. Important European advances can be expected in near future. Amongst all noise reducing measures, noise barriers placed close to the source are probably best known. Unfortunately, it is often difficult to apply them in urban areas due to space restrictions and visual impact.

The Ideal Urban Sonic Environment

A naive view of urban soundscape design may aim at making the sonic environment as quiet as possible. In a more modern view, absolute quietness is believed not to be a necessity and in some circumstances may even be unwanted. The ideal urban sonic environment depends on the context.

Dwellings: Home Environment

At home, extensive environmental mechanical or electronic noise is perceived as intruding the observer's private space. As such it will be perceived as disturbing and annoying as soon as it is noticed. This immediately implies that different types of traffic noise will result in different annoyance for the same energetically averaged exposure. Today, the relationships between façade exposure measured in L_{den} and self-reported long term annoyance have been well established for different types of traffic noise and are quite generally accepted. These curves show that for the same average exposure, annoyance seems to be greatest for aircraft noise and lowest for train noise. For industrial noise, characteristics of the sound may differ significantly between sites and these characteristics (tonality, impulse) may increase noticeability and thus annoyance. This has been taken into account in regulation such as Vlarem 2 that are applicable for industrial noise.

Assessing environmental noise levels at the façade is much easier than measuring noise at the actual ear of the urban dweller and hence most scientific advances in relating exposure to effects have been made in that area as stated above. It should nevertheless be remembered that what matters is the sound that reaches the ear. The dwelling itself acts as a shield for the intruding noise - and in some particular cases as an amplifier for low frequency sound. At the one hand the dwelling may (over)amplify low frequency problems, at the other hand it creates opportunities for indoor noise reduction. Noise reduction by increased building insulation requires closing all windows and doors at any time and thus may limit the sense of control and freedom of the inhabitants which in turn has a negative effect on perceived annoyance. This has led to the observation that insulation requirements should be relaxed at least on one façade and this in turn led to the introduction of the concept: quiet side. Availability of a silent or highly shielded façade relaxes annoyance by an equivalent of a reduction of most exposed façade level by 5 to 10 dBA as long as the level at the quiet side does not exceed 55 dBA, and the difference between the loudest and quiet side is at least 10 dB. Furthermore, the noise levels at the mostexposed façade may not be excessively high. There is however also growing evidence that the sonic environment in the wider neighborhood of the house contributes to the annoyance in particular for people living in apartments.

Disturbance of activity and in particular sleep disturbance are important as a first stage negative effect of intruding environmental noise. Subjective evaluation of sleep quality and fatigue correlate well with energetically averaged nightly noise levels, $L_{Aeq,night}$, but other sleep indicators (changes in heart rate, sleep stage, etc.) show a more complicated relationship with number and level of noise events. There is some evidence that habituation to traffic noise events during the night does not occur. It has also been established that noise events occurring during the end of sleep have a stronger effect on overall sleep quality and thus traffic curfews are most effective in the morning. Disturbances occurring during the beginning of the night seem to be compensated. Concerning sleep disturbance, no clear difference between different types of traffic could be established indicating that the meaning of sound (within the scope of traffic noises) may not be as important during sleep as could be expected.

Health effects occur at much longer time scales and thus the analyses and design of the acoustic environment may need less temporal detail. Today the best proven effect of long exposure to environmental noise in the home environment is ischemic heart disease, which may in turn be related to blood pressure. The effect threshold seems to lie around 65 dBA (daytime equivalent noise level in front of the façade) but odds ratios are small, even 10 dBA above that threshold. Since it is estimated that the causal path between exposure and health effects of environmental noise involves the cognitive/emotional path for 75%, urban noise management aiming at reducing annoyance can be expected also to reduce health effects.

Urban Public Space

City dwellers perceive and evaluate the soundscape of the urban public space in a rather different way than they perceive the intruding sounds in their home environment. Thus, modern urban soundscape planning is evolving new ideas and concepts to accommodate this difference. The soundscape is seen as an integral part of the urban environment, contributing to the identify and specificity of this environment. The quality of a soundscape is assessed within the particular context and use imposed by the urban space. The physical characteristics of the sonic environment needed to evaluate this quality go far beyond the overall noise level and include spectral and temporal structure. Today the quest for appropriate physical indicators is continuing but one could argue that physical indicators will never be sufficient and the meaning that the listener associates to the sound is most important.

The urban public space has different uses: shopping, moving between functions, recreation. The soundscape of parks and squares mainly used for recreation has often been in focus. An important reason for this is that the availability of high quality (green, quiet) urban space within reach was proven to benefit the health of urban dwellers. In particular, the potential for psychological restoration of natural quiet areas has been suggested. Thus the goal for urban planning in relation to traffic noise is clearly broadened.

Urban Planning and Development

Urban planning and development has long discarded noise as a point of concern. It was believed that environmental noise issues could best be tackled after all geographic and visual planning of the city had been finished or in the best case at the very end of the planning stage. Most technical measures at the source are indeed still applicable then, but they rarely are efficient in urban context. In the new perspective, sound is an integral part of the urban setting and thus should be considered at the same level of importance as visual esthetics.

Living Areas with Low Exposure

Noise levels in general decrease quite rapidly as the distance to the source is increased. In open area, near the source, noise levels drop by 6 dBA with doubling of distance when the source is a well localized point and by 3 dBA with doubling of distance for a line source (a straight road for example). In natural areas and parks, the ground acts as an acoustical porous material. Sound propagating above such a material is strongly attenuated over a certain frequency range. This so called ground effect can result in a significant additional noise reduction. It does however require that sound shears the ground and thus the positive effect vanishes under several conditions: when strong temperature inversion or a wind gradient bend the sound waves downward; when the sound propagates over a valley; when the sound source is elevated for example on a bridge; when the observer is elevated, for example living in a high rise building. In the urban context, screening by buildings can reduce sound levels considerably thus leading to a decrease with distance that is well above the 3 or 6 dBA rule mentioned above. This is why it is so important to increase the distance between urban dwellings or recreational areas and the source of noise: traffic. To take advantage of the additional ground effect and screening by buildings, traffic routes are best planned as low as possible within the 3-dimensional city.

In many Flemish cities and villages, road traffic is at the level of the surrounding terrain and houses form a continuous screen along this sound source. This explains why the percentage of highly noise annoyed people drops considerably for houses somewhat further away from the main roads. In the a real photograph of part of a city and the corresponding noise map can be seen that some zoning occurs naturally because of commerce and SME moving to the arterial road but also that it is far from perfect.

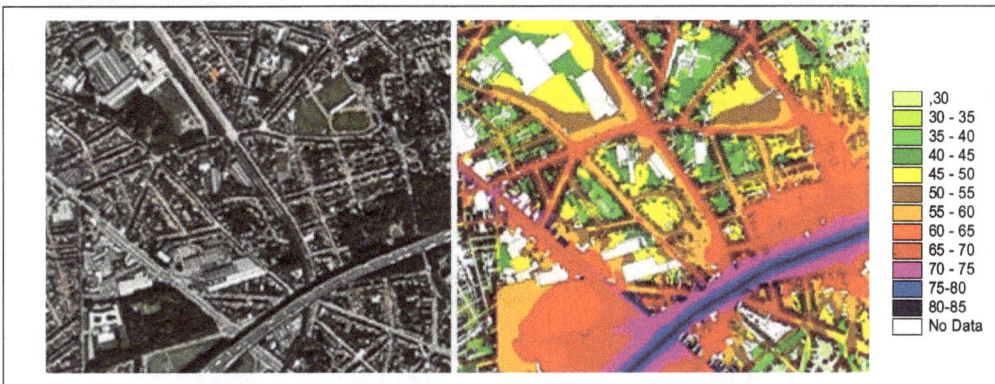

A real photograph (left) and corresponding noise map (right) for part of a city.

Noise Barrier Buildings and Quiet Sides

Silent façades can be achieved in various ways. A cluster of buildings, with a central courtyard is an interesting configuration, and provides a large number of buildings with a quiet side. A "noise barrier building" has both a quiet side and moderate noise levels behind the most-exposed façade. Such a façade only has a limited number of (acoustically highly insulated) windows. The silent (or noise-sensitive) side on the other hand contains the necessary windows and doors, balconies and gardens. Very efficient noise barrier buildings could make use of a climbing earth berm against the most exposed-side. When noise barrier buildings are connected together, their efficiency is largely increases because side diffraction is prevented.

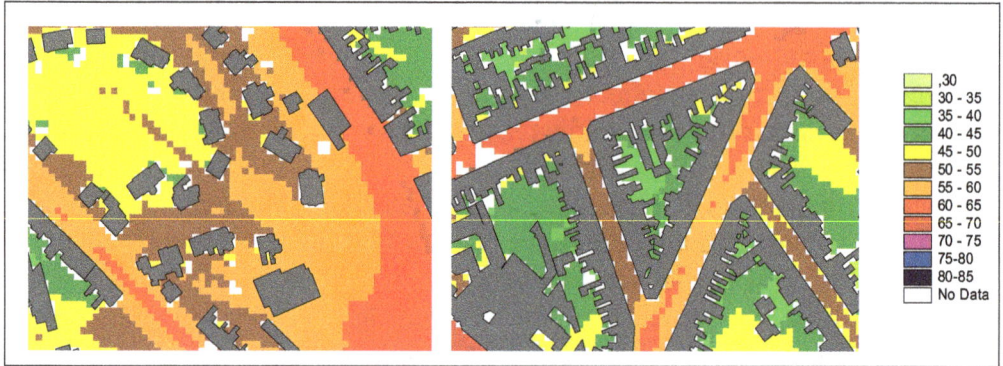

Two excerpts from a city noise map (Lden) showing building structure with pronounced quiet side (left) and buildings where a quiet side is almost absent (right).

Vegetated roofs tops (green roofs) can help in achieving quiet façades. The substrates used for both extensive (mostly granular material) and intensive (uncompacted earth) green roofs have sound absorbing properties. Sound diffracting over the roof will be attenuated more than when it propagates over classical, acoustical hard roof coverage. Numerical simulations showed the potential of using green roofs to reduce the noise impact near buildings for example in the situation shown in Figure. Important parameters in this respect are the sound frequency, the layer thickness of the substrate, the building geometry, and traffic related parameters like vehicle speed and vehicle type. Besides reducing sound waves shearing over the building, a reduction in the transmission of sound through the roof construction is obtained. More generally, roof type and roof slope angle should be considered when the sound pressure levels at non-directly exposed façades are of interest.

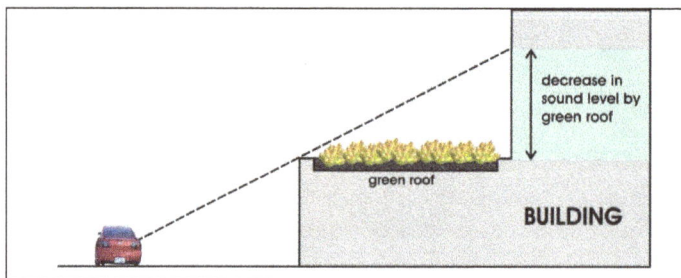

Example of a building geometry leading to a decrease in sound level at part of the facade caused by the presence of a green roof.

Street Reverberation

Environmental noise heard in an environment with a long reverberation time is often perceived as very annoying. A so-called street canyon, i.e. a narrow street, enclosed by tall and connected buildings, induces such long reverberation times. Such street geometries are typically observed in the centers of (historically grown) European cities. Due to the confinement of the sound in the street, the sound decay by geometrical spreading is low. The road surface, the footway, and façades of buildings mainly consist of acoustically rigid materials. This absence of absorbing materials further increases the reverberation.

Besides specular reflection (also called mirror source reflection), a non-flat surface also induces (to some degree) diffuse reflection. As a result, the incident acoustical energy is spread over a range of directions, while sound reflection from a fully flat surface is very directive. Architectural ornaments, window sills, and protrusions and recessions by windows increase diffuse reflection in a street. Large surfaces of glass are known as specular elements. In case of diffuse reflection, part of the acoustical energy is also reflected in upward and sideward direction. This allows sound to leave the street canyon already after a limited number of diffuse reflections which causes noise levels to drop and reverberation time to shorten.

Increasing the absorption of façades largely reduces sound pressure levels in the street. The (classical) porous absorbing materials are often not suited for application near the façades because they are not weather-resistant. A possible application of such materials, however, is at the underside of balconies. Vegetation near facades is another interesting option. By means of new techniques it is possible to fix the necessary substrates at a few centimeters from the walls. Given the large number of reflections between the façades in a street canyon, and since such substrate are highly porous, strong reductions of noise levels and street reverberation may be achieved. Reducing extensive reflections in the street canyon also reduces noise levels at the least exposed façade.

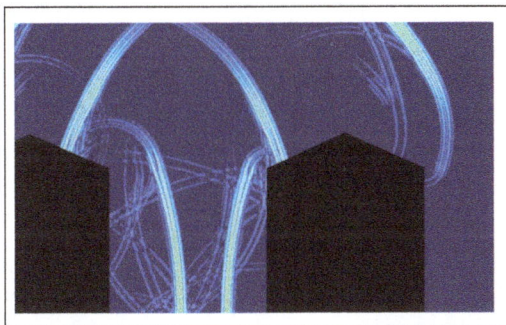

Wavefronts travelling back and forth in a street canyon after a short acoustic pulse is emitted in the center of the canyon.

Urban Open Space

Traditional noise control engineering has two main disadvantages when it is solely applied in order to mitigate noise at more quiet urban areas with recreational purposes, such as urban squares and parks. Firstly, the traditional approach may result in a greying of the sonic environment, because often black spots are targeted, at a disadvantage of the sound pressure level at other places. Secondly and more importantly, noise control engineering is a negative approach: not all sounds are noise, some sounds do fit well in some environments, and we should strive to preserve these sounds rather than to eliminate them. A more positive and holistic approach is needed, aimed at designing entire environments that are pleasing to the ear.

Soundscape Description

A high quality sonic environment can be defined as a sonic environment in which there is a good match between the sounds that can be heard (commonly referred to as the soundscape, as an analogy to the term landscape), and the sounds that are expected. In other words, a high quality soundscape contains lots of fitting sounds that can be clearly heard, and less non-fitting sounds. Unfortunately, defining which sounds fit in a given environment is a complex and interdisciplinary problem. Understanding the factors which influence the perception of environmental sounds forms the main subject of acoustic ecology, which is the study of the interactions, mediated through sound, between humans and their environment. Whereas traditional noise control engineering solely involves physical measures, acoustic ecology departs from a human-centered viewpoint.

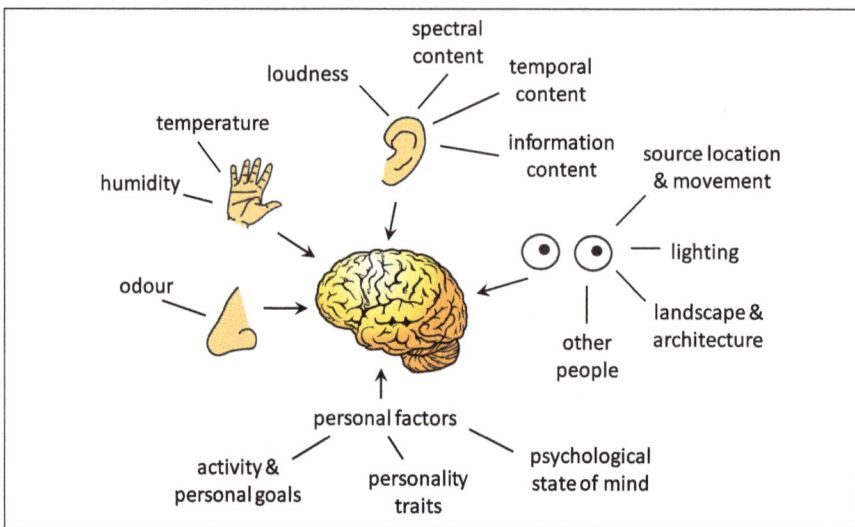

Various factors in the perception of environmental sounds.

The perception of sounds, and as such their perceived degree of fit to the environment, is determined by both sensory and personal factors. Sensory factors include auditory aspects, such as the loudness, spectral, temporal and information content of the sound, visual aspects such as the location and movement of the source (if it is visible), the landscape and architecture, lightning, activities of other people, tactile aspects such as temperature and humidity, and olfactory aspects. Personal factors include traits such as noise sensitivity and attitude towards different types of sources, the current activity and personal goals, and the current emotional state. These various factors are visualized in figure.

Planning and Acoustic Design

The visual aspect has been, up to now, the most important factor in the design of urban parks and open spaces. However, including auditory aspects and knowledge on perception of soundscapes in urban planning and design, an approach often referred to as acoustic design, has great potential.

Urban public spaces can be designed to encourage activities which generate unique sounds or soundmarks that attract attention and reflect traditional or cultural elements. A first example is music. Studies suggest that the low frequency content in live music is often not loud enough to mask traffic sound. High frequency components, on the other hand, can make the music stand out from the background, making the soundscape more pleasant. Another example is the sound from water fountains. Altering flow methods and fountain design has proven to provide great potential in shaping the spectrum of water features, making it an ideal instrument for attracting attention and masking traffic noise. Adding greenery in well-arranged spaces may enhance the natural feeling of the environment and alter the sound pressure level distribution, but may also attract songbirds. As a more drastic measure, (camouflaged) loudspeakers can be introduced into the design, which could play back fitting environmental sounds, such as singing birds in an urban park.

Auralisation forms an important tool in acoustic design. This technique aims at a realistic, artificial simulation and reproduction of the various sound sources that can be heard in a given environment, such as traffic, fountains, street music, human voices etc., the full path that sounds travel, from emission at the source to reception at the ear, is hereby modeled. Reflections and diffractions of sound on objects have to be taken into account, as well as the Doppler Effect for moving sources or listeners. In order to achieve a realistic representation, the auralisation should be accompanied with a (3D) visual representation of the virtual environment.

Artificial soundscapes produced with the auralisation technique have still to be listened to by human listeners in order to be able to assess their quality, which limits the applicability of this approach. In the future, models for automatic acoustic evaluation could replace the human listener. Several approaches have already been suggested, such as artificial neural network based models, or approaches that try to model the human perception of sound in a bottom-up fashion, starting from basic psychoacoustic and psychological principles.

Assessment of Quiet Areas

Generally, a quiet area is defined as an area that is quieter than the surrounding region, and which has a psychological restoring effect on people visiting it. There is a growing awareness that quiet areas deserve special attention and preservation, and this goal has therefore been subscribed in the Environmental Noise Directive of the European Commission and in policy intentions of many countries. In line with the ideas described above, an (urban) quiet area such as a park or open space does not imply the absence of sound (which would be silence). Rather, its soundscape should be experienced as quiet by the average visitor. Quality assessment methods for urban quiet areas have to reflect this principle; the average sound pressure level is therefore less suited as the only indicator to characterize quiet areas. Moreover, one should go beyond the use of only quantitative approaches.

A starting point could be to consider the soundscape of urban quiet areas as the superposition of an always present background and sound events. This subdivision is illustrated in Figure. The background largely determines the overall feeling of quietness, and thus the basic quality of the soundscape. It can be heard, but it does not trigger much meaning because it is not listened to consciously. Events can disturb the soundscape, but it is also possible that they, being fitting sounds, accentuate the basic quality. It is known that the perception of sound events involves source recognition and association. The background, on the other hand, may not lead to any source recognition; it may be experienced in a more holistic way.

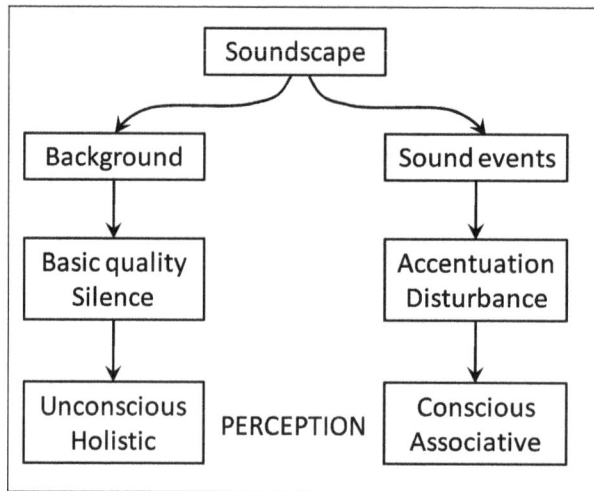

Various aspects related to the urban soundscape.

A multicriteria approach to the quality assessment of urban quiet areas has to address the quality of both background and sound events, and has to include the different perceptual factors. Based on a meta-analysis of several studies on soundscape perception, the authors have proposed a quality assessment methodology consisting of the following subjective and objective criteria: pleasantness and the presence and number of non-fitting sound events, determined using a questionnaire survey; quality of the background, measured using indicators for loudness, temporal and spectral content; congruence of the area; biological, natural and landscape value of the environment. The proposed set of indicators forms a balance between scientific validity, applicability and comprehensibility.

Orchestrating Traffic Noise

Reducing road traffic noise at the source is more than reducing engine noise and improving tire-road interaction. By modifying the traffic stream itself, through careful traffic planning, many aspects of the urban sonic environment can be tuned. Dynamics and temporal structure depend on the pass by of individual cars, trucks, and motorcycles. Microscopic management can help designing this detail of the soundscape. On a wider spatial horizon, reorganizing traffic streams can be very useful.

Dynamics and Temporal Structure: Microscopic Management

Strategic, large scale traffic management decisions, for example on the scale of a city, are most often based on estimates of average traffic flow through main roads. However, over the past decades, the awareness has grown among traffic researchers that small-scale changes in infrastructure, and even in driving behavior of individual vehicles, can have a large influence on (macroscopic) traffic flow. A good example is the green wave induced by the coordinated use of traffic lights at several successive signalized intersections along a driving direction of a road. Also stimulated by the growing availability of computing power, traffic researchers and engineers are therefore more and more considering the use of microscopic simulation models in their study of urban mobility problems.

Microscopic models (micromodels in short) consider the exact location and movement of individual vehicles over time, within an (urban) environment that is modeled in high detail (locations of kerbs and stop lines, exact size of intersections etc). Behavior rules, such as the distance to keep to the vehicle in front or when to change lanes, form the core of the model. Micromodels allow traffic engineers to gain insight in the course of complex phenomena such as the formation of jams or the propagation of traffic density waves.

Micromodels have great potential as a tool in acoustic design of the urban soundscape. When coupled with a noise emission model for single vehicles and a detailed propagation model, micro models allow estimating the (temporal structure of) peaks in the sound pressure level caused by vehicle pass-bys for an example of this approach). Figure shows an example of the measured time-varying sound pressure level at the kerbside of the Frederik Burvenichstraat in Gentbrugge, Belgium, together with a simulated time series at the same spot within a micromodel of the city of Gentbrugge. Note that there is only a statistical similarity, not an exact one, because of the stochastic nature of microsimulation models. Nevertheless, this approach allows estimating peak levels and statistical levels such as the median sound pressure level with good accuracy. As such, the influence of detailed traffic management measures, such as speed bumps, roundabouts or speed control and of sound propagation measures such as inserting noise barriers, on the temporal structure of the urban soundscape can be assessed.

More detailed study of auditory perception of car and truck passages allowed to estimate the subjective annoyance (used as an opposite to sound quality in this study) on the basis of specialized sound quality measures: relative approach, peak loudness (5 percentile), and peak sharpness (5 percentile). Relative approach measures abrupt changes in time of frequency. This sound quality approach is now being extended to traffic streams as a whole. Micromodels create an unsurpassed potential for tuning this sound quality using different traffic measures.

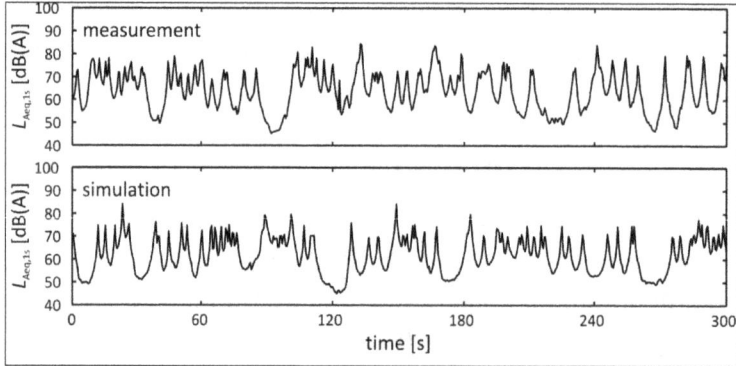

Measured and simulated 5-minute time series of the sound pressure level caused by traffic noise at the kerbside of the Frederik Burvenichstraat in Gentbrugge during rush hour.

Next to their use in modeling the time-varying sound pressure level in urban environment, micromodels can also be applied in the assessment of the impact of dynamic vehicle parameters, such as acceleration, on the average sound pressure level. Urban intersections are an obvious point of interest, because of the typical acceleration and deceleration pattern of traffic near intersections. Classical traffic noise estimation models based on average flows do not allow taking into account the acceleration or deceleration of vehicles correctly, but they can be corrected based on more detailed microsimulation results of traffic at and near intersections.

Calming Urban Traffic: Macroscopic Management

Calming urban traffic in certain parts of the city is an alternative for optimally locating dwellings: if you cannot move people away from traffic, move traffic away from the living areas. This is not always as straight forward as it seems at first sight. A drastic halving of traffic intensity theoretically reduces overall noise levels by 3 dBA, but when taking into account that the traffic will move more freely, hence faster, one may end up with virtually no noise reduction. Depending on the strategy for implementing the reduction of traffic intensity, local fleet composition may change (e.g. more public busses due to private traffic charging). If traffic calming measures are applied to particular road segments and/ or particular parts of the day (night ban, congestion charging) this may lead to traffic increases on alternative roads and at alternative times of the day. Careful traffic modeling with additional focus on day, evening and night periods is needed to distinguish between noise reduction and noise redistribution. Unfortunately, current traffic modeling exercises often neglect nightly traffic or only make rough assumptions about it.

To quantify these observations, we present a few examples of simulated traffic noise emission in this section. Three prototype roads are considered: a highway (HW) of three lanes, a 2 by 2 lanes major road (MR) with divided directions and a local road (LR) with 1 lane for each direction. In addition, the three fleet compositions shown in table are considered: Fleet 1 corresponds to the current situation in Belgium, Fleet 2 strongly promotes alternative fuels and Fleet 3 bans petrol and diesel from cars all together. This alternative fleet may seem rather unrealistic, but one should consider that they could correspond to the local situation in parts of a city. Fleet 3 for example could correspond to part of a city where classical fuel cars are strongly disencouraged. Still, implementing such drastic scenarios would take several decades.

Table: Fleet compositions used in the model.

	Fuel type	petrol	Diesel	LPG	CNG	Hybrid	Electric
Fleet 1	Cars	44.2%	54.5%	0.9%	0	0.4%	0
	Light	44.2%	54.5%	0.9%	0	0.4%	0
	Heavy	0	100%	0	0	0	0
Fleet 2	Cars	20.0%	20.0%	0	10.0%	25.0%	25.0%
	Light	20.0%	20.0%	0	10.0%	25.0%	25.0%
	Heavy	0	80.0%	0	0	20.0%	0
Fleet 3	Cars	0	0	0	25.0%	25.0%	50.0%
	Light	0	0	0	25.0%	25.0%	50.0%
	Heavy	0	50.0%	0	0	50.0%	0

Noise emission per vehicle is calculated based on the Harmonoise/Imagine road traffic source model using traffic volume and speed as the main parameters. The emission data are tuned to account for quieter propulsion and corresponding lowering of rolling noise in hybrid and electric cars. The latter is a positive side effect: it was observed that car manufacturers opt for quiet tires on their hybrid cars. Capacity is estimated at HW:4590 veh/h, MR:2430 veh/h, and LR:1125 veh/h and average speed is deducted from speed-capacity relationships counting one heavy vehicle as two cars. Average speed reduction due to congestion is entered in the emission model rather than a more realistic speed distribution and acceleration or deceleration that can be obtained from micro-simulation. Simulation results are shown in Figure.

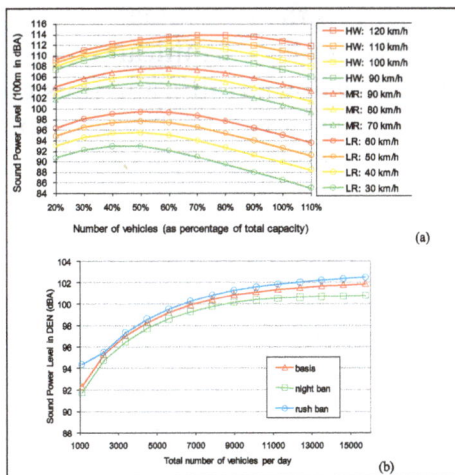

Sound power level emitted by 100m of road (one driving direction) as a function of traffic volume: for different driving speed (a), for different fleets on highway and main road (c) and on local road (d), and for two scenarios of heavy traffic ban (b, including day-evening-night weighting).

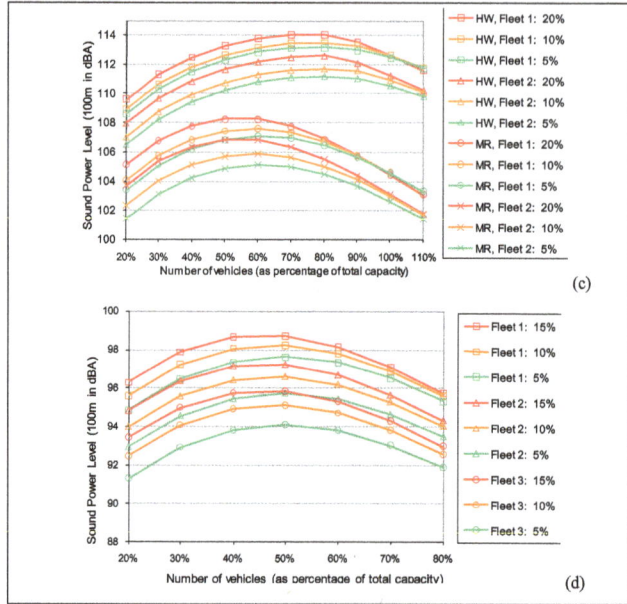

From figure the significant effect of speed limit on noise emission can be recognized. Note that the traffic volume is expressed relative to the capacity of the different road prototypes and that the percentage of heavy traffic differs amongst road types (HW: 15%, MR: 10%, LR: 5%). Road saturation and corresponding lowering of traffic speed reduces noise emission. This saturation effect was validated against long term measurements along a Flemish highway in. The rather flat top of the curves explains why reducing traffic volumes not always leads to expected sound level reduction.

The influence of fleet composition on noise emission is shown in figure for highway (speed limit 120km/h) and major road (speed limit 90km/ h) and in figure for local roads (speed limit 50km/h). As the percentage of heavy traffic increases, noise emission increases since trucks produce significantly more noise than cars. Closer to saturation, trucks tend to slow down traffic and thus the overall noise emission depends less on percentage of heavy traffic. The influence of rather drastic fleet changes is significant, but comparable to effects of speed limit and percentage of heavy goods traffic.

Cost Efficiency Issues

Cost-benefit Analysis

In the derivation of noise standards from noise guidelines two different approaches for decision making can be applied. Decisions can be based purely on health, cultural and environmental consequences, with little weight to economic efficiency. This approach has the objective of reducing the risk of adverse noise effects to a socially acceptable level. The second approach is based on a formal cost-effectiveness, or cost-benefit analysis (CBA). The objective is to identify control actions that achieve the greatest net economic benefit, or are the most economically efficient. The development of noise standards should account for both extremes, and involve stakeholders and assure social equity to all the parties involved. It should also provide sufficient information to guarantee that stakeholders understand the scientific and economic consequences.

To determine the costs of control action, the abatement measures used to reduce emissions must be known. This is usually the case for direct measures at the source and these measures can be monetarized. Costs of action should include all costs of investment, operation and maintenance. It may not be possible to monetarize indirect measures, such as alternative traffic plans or change in behaviour of individuals.

The steps in a cost-benefit analysis include:

- The identification and cost analysis of control action (such as emission abatement strategies and tactics).

- An assessment of noise and population exposure, with and without the control action.

- The identification of benefit categories, such as improved health and reduced property loss.

- A comparison of the health effects, with and without control action.

- A comparison of the estimated costs of control action with the benefits that accrue from such action.

- A sensitivity and uncertainty analysis.

Action taken to reduce one pollutant may increase or decrease the concentration of other pollutants. These additional effects should be considered, as well as pollutant interactions that may lead to double counting of costs or benefits, or to disregarding some costly but necessary action. Due to different levels of knowledge about the costs of control action and health effects, there is a tendency to overestimate the cost of control action and underestimate the benefits.

CBA is a highly interdisciplinary task. Appropriately applied, it is a legitimate and useful way of providing information for managers who must make decisions that impact health. CBA is also an appropriate tool for drawing the attention of politicians to the benefits of noise control. In any case, however, a CBA should be peer-reviewed and never be used as the sole and overriding determinant of decisions.

Noise Control

Noise Control Engineering

As with any occupational hazard, control technology should aim at reducing noise to acceptable levels by action on the work environment. Such action involves the implementation of any measure that will reduce noise being generated, and will reduce the noise transmission through the air or through the structure of the workplace. Such measures include modifications of the machinery, the workplace operations, and the layout of the workroom. In fact, the best approach for noise hazard control in the work environment is to eliminate or reduce the hazard at its source of generation, either by direct action on the source or by its confinement.

Practical considerations must not be overlooked; it is often unfeasible to implement a global control program all at once. The most urgent problems have to be solved first; priorities have to be set up. In certain cases, the solution may be found in a combination of measures which by themselves would not be enough; for example, to achieve part of the required reduction through environmental measures and to complement them with personal measures (e.g. wearing hearing protection for only 2-3 hours), bearing in mind that it is extremely difficult to make sure that hearing protection is properly fitted and properly worn.

This topic presents the principles of engineering control of noise, specific control measures and some examples. Many of the specific noise control measures described are intended as a rough guide only.

Noise Control Strategies

Prior to the selection and design of control measures, noise sources must be identified and the noise produced must be carefully evaluated.

To adequately define the noise problem and set a good basis for the control strategy, the following factors should be considered:

- Type of noise.

- Noise levels and temporal pattern.

- Frequency distribution.

- Noise sources (location, power, and directivity).

- Noise propagation pathways, through air or through structure.

- Room acoustics (reverberation).

In addition, other factors have to be considered; for example, number of exposed workers, type of work, etc. If one or two workers are exposed, expensive engineering measures may not be the most adequate solution and other control options should be considered; for example, a combination of personal protection and limitation of exposure.

The need for control or otherwise in a particular situation is determined by evaluating noise levels at noisy locations in a facility where personnel spend time. If the amount of time spent in noisy locations by individual workers is only a fraction of their working day, then local regulations may allow slightly higher noise levels to exist. Where possible, noise levels should be evaluated at locations occupied by workers' ears.

Normally the noise control program will be started using as a basis A-weighted immission or noise exposure levels for which the standard ISO 11690-1 recommends target values and the principles of noise control planning. A more precise way is to use immission and emission values in frequency bands as follows.

The desired (least annoying) octave band frequency spectrum for which to aim at the location of the exposed worker is shown in figure for an overall level of 90 dB(A). If the desired level after control is 85 dB(A), then the entire curve should be displaced downwards by 5 dB. The curve is used by determining the spectrum levels in octave bands and plotting the results on the graph to determine the required decibel reductions for each octave band. Clearly it will often be difficult to achieve the desired noise spectrum, but at least it provides a goal for which to aim.

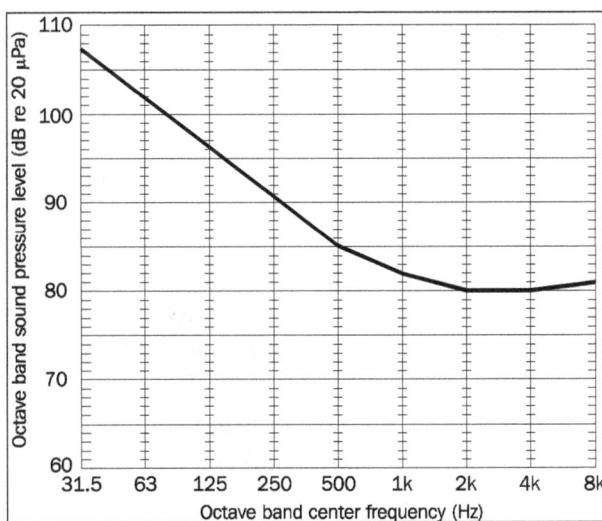

Desired noise spectrum for an overall level of 90 dB(A).

It should be noted that because of the way individual octave band levels are added logarithmically, an excess level in one octave band will not be compensated by a similar decrease in another band. The overall A-weighted sound level due to the combined contributions in each octave band is obtained by using the decibel addition procedure.

Any noise problem may be described in terms of a source, a transmission path and a receiver (in this context, a worker) and noise control may take the form of altering any one or all of these elements. The noise source is where the vibratory mechanical energy originates, as a result of a physical phenomenon, such as mechanical shock, impacts, friction or turbulent airflow. With regard to the noise produced by a particular machine or process, experience strongly suggests that when control takes the form of understanding the noise-producing mechanism and changing it to produce a quieter process, as opposed to the use of a barrier for control of the transmission path, the unit cost per decibel reduction is of the order of one tenth of the latter cost. Clearly, the best controls are those implemented in the original design. It has also been found that when noise control is considered in the initial design of a new machine, advantages manifest themselves resulting in a better machine overall. These unexpected advantages then provide the economic incentive for implementation, and noise control becomes an incidental benefit. Unfortunately, in most industries, occupational hygienists are seldom in the position of being able to make fundamental design changes to noisy equipment. They must often make do with what they are supplied, and learn to use effective "add-on" noise control technology, which generally involves either modification of the transmission path or the receiver, and sometimes the source.

If noise cannot be controlled to an acceptable level at the source, attempts should then be made to control it at some point during its propagation path; that is, the path along which the sound energy from the source travels. In fact, there may be a multiplicity of paths, both in air and in solid structures. The total path, which contains all possible avenues along which noise may reach the ear, has to be considered.

As a last resort, or as a complement to the environmental measures, the noise control problem may be approached at the level of the receiver, in the context of this document, the exposed worker(s).

In existing facilities, controls may be required in response to specific complaints from within the workplace, and excessive noise levels may be quantified by suitable measurements as described previously. In proposed new installations, possible complaints must be anticipated, and expected excessive noise levels must be estimated by some procedure. As it is not possible to entirely eliminate unwanted noise, minimum acceptable levels of noise must be formulated and these levels constitute the criteria for acceptability which are generally established with reference to appropriate regulations in the workplace.

In both existing and proposed new installations an important part of the process is to identify noise sources and to rank order them in terms of contributions to excessive noise. When the requirements for noise control have been quantified, and sources identified and ranked, it is possible to consider various options for control and finally to determine the cost effectiveness of the various options. As was mentioned earlier, the cost of enclosing a noise source is generally much greater than modifying the source or process producing the noise. Thus an argument, based upon cost effectiveness, is provided for extending the process of source identification to specific sources on a particular item of equipment and rank ordering these contributions to the limits of practicality.

Existing Installations and Facilities

In existing facilities, quantification of the noise problem involves identification of the source or sources, determination of the transmission paths from the sources to the receivers, rank ordering of the various contributors to the problem and finally determination of acceptable solutions.

To begin, noise levels must be determined at the locations from which the complaints arise. Once levels have been determined, the next step is to apply acceptable noise level criteria to each location and thus to determine the required noise reductions, generally as a function of octave or one-third octave frequency bands.

Once the noise levels have been measured and the required reductions determined, the next step is to identify and rank order the noise sources responsible for the excessive noise. The sources may be subtle or alternatively many, in which case rank ordering may be as important as identification. Where many sources exist, rank ordering may pose a difficult problem.

When there are many sources it is important to determine the sound power and directivity of each to determine their relative contributions to the noise problem. The radiated sound power and directivity of sources can be determined by reference to the equipment manufacturer's data or by measurement. The sound power should be characterised in octave or one third octave frequency bands and dominant single frequencies should be identified. Any background noise interfering with the sound power measurements must be taken into account and removed.

This is the ideal procedure. In reality, many people choose machinery or equipment using only noise emission values according to ISO 4871 and they make comparisons according to ISO 11689.

Often noise sources are either vibrating surfaces or unsteady fluid flow (air, gas or steam). The latter are referred to as aerodynamic sources and they are often associated with exhausts. In most cases, it is worthwhile to determine the source of the energy which is causing the structure or the aerodynamic source to radiate sound, as control may best start there.

Having identified the noise sources and determined their radiated sound power levels, the next task is to determine the relative contribution of each noise source to the noise level at each location where the measured noise levels are excessive. For a facility involving just a few noise sources, as is the case for most occupational noise problems at a specific location, this is usually a relatively straightforward task.

Once the noise sources have been ranked in order of importance in terms of their contribution to the overall noise problem, it is often also useful to rank them in terms of which are easiest to do something about and which affect most people, and take this into account when deciding which sources to treat first of all.

Installations and Facilities in the Design Stage

In new installations, quantification of the noise problem at the design stage may range from simple to difficult but never impossible. At the design stage the problems are the same as for existing installations; they are identification of the source or sources, determination of the transmission paths of the noise from the sources to the receivers, rank ordering of the various contributors to the problem and finally determination of acceptable solutions. Most importantly, at the design stage the options for noise control are generally many and may include rejection of the proposed design.

The first step for new installations is to determine the noise criteria for sensitive locations which may typically include locations of operators of noisy machinery. If the estimated noise levels at any sensitive location exceed the established criteria, then the equipment contributing most to the excess levels should be targeted for noise control, which could take the form of:

- Specifying lower equipment noise levels to the equipment manufacturer (care must be taken whenever importing equipment, particularly second hand which can be very noisy and hence no longer acceptable in the country of origin);

- Including noise control fixtures (mufflers, barriers, vibration isolation systems, enclosures, or factory walls with a higher sound transmission loss) in the factory design;

- Rearrangement and careful planning of buildings and equipment within them. The essence of the discussion is that sources placed near hard reflective surfaces will result in higher sound levels at the approximate rate of 3 dB for each large surface, as illustrated in figure. Note that the shape of the building space generally is not important, as a reverberant field can build-up in spaces of any shape. Care should be taken to organise production lines so that noisy equipment is separated from workers as much as possible.

Sufficient noise control should be specified to leave no doubt that the noise criteria will be met at every sensitive location. Saving money at this stage is not cost effective in the long term.

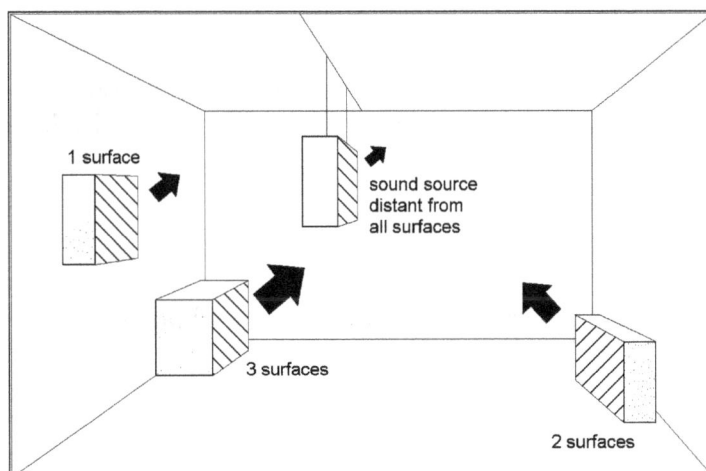

Sound sources should not be placed near corners.

Control of Noise at the Source

To control noise at the source, it is first necessary to determine the cause of the noise and secondly to decide on what can be done to reduce it. Modification of the energy source to reduce the noise generated often provides the best means of noise control. For example, where impacts are involved, as in punch presses, any reduction of the peak impact force (even at the expense of a longer time period over which the force acts) will dramatically reduce the noise generated.

Generally, when a choice of mechanical processes is possible to accomplish a given task, the best choice, from the point of view of minimum noise, will be the process which minimises the time rate of change of force or jerk (time rate of change of acceleration). Alternatively, when the process is aerodynamic a similar principle applies; that is, the process which minimises pressure gradients will produce minimum noise. In general, whether a process is mechanical or fluid mechanical, minimum rate of change of force is associated with minimum noise.

Among the physical phenomena which can give origin to noise, the following can be mentioned:

- Mechanical shock between solids.
- Unbalanced rotating equipment.
- Friction between metal parts.
- Vibration of large plates.
- Irregular fluid flow, etc.

Control of noise at the source may be done either indirectly, i.e. generally, or directly, i.e. related to the design process addressing one of the causes.

Noise Control Techniques

The following are noise control techniques that will produce substantial noise reductions quickly, with little or no effect on normal operation or use in the industry.

Damping

Typical Applications

Chutes, hoppers, machine guards, panels, conveyors and tanks.

Technique

There are two basic techniques:

- Unconstrained layer damping where a layer of bitumastic (or similar) high damping material is stuck to the surface.

- Constrained layer damping where a laminate is constructed.

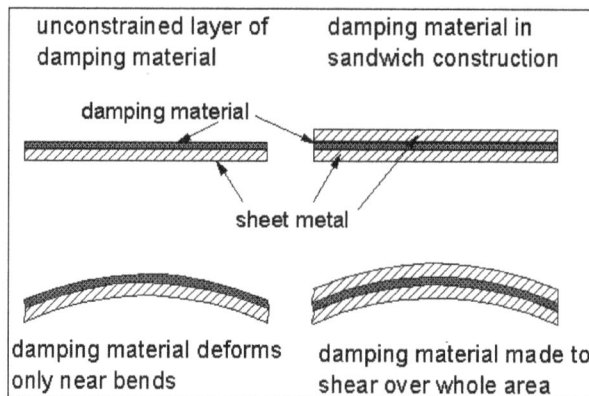

Constrained layer damping is more rugged and generally more effective. Either remanufacture steel (or aluminium) guards, panels or other components from commercially available sound deadened steel or buy self-adhesive steel sheet. The latter can simply be stuck on to existing components (inside or outside) covering about 80% of the flat surface area to give a 5 - 25 dB reduction in the noise radiated (use a thickness that is 40% to 100% of the thickness of the panel to be treated).

Limitations: The efficiency falls off for thicker sheets. Above about 3mm sheet thickness it becomes increasingly difficult to achieve a substantial noise reduction.

Fan Installations

Typical Applications

Axial flow or centrifugal fans.

Technique

Maximum fan efficiency coincides precisely with minimum noise. Any fan installation feature that tends to reduce fan efficiency is therefore likely to increase noise. Two of the most common examples are bends close to the fan (intake side in particular) and dampers (close to the fan intake or exhaust).

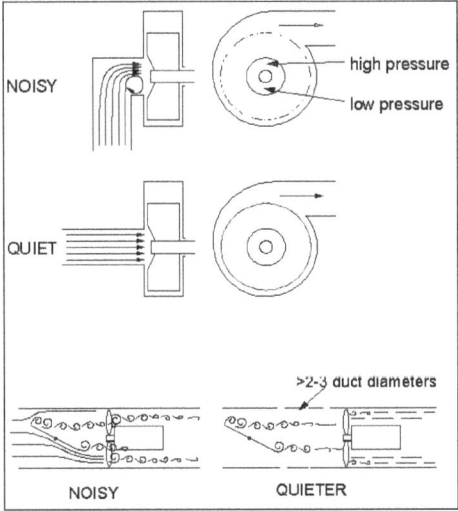

Ideally, for maximum fan efficiency and minimum noise, make sure there is at least 2 - 3 duct diameters of straight duct between any feature that may disturb the flow and the fan itself. Noise reductions of 3 - 12 dB are often possible.

Ductwork

Typical Applications

Extraction, ventilation, cooling, openings in walls and enclosures.

Technique

Instead of fitting silencers, it is often possible to achieve a 10 - 20 dB reduction in airborne noise from a duct or opening by lining the last bend in the ductwork with acoustic absorbent (foam or rockwool/fibreglass). Alternatively, construct a simple absorbent lined right-angled bend to fit on the opening. Ideally, either side of the bend should be lined along a length equivalent to twice the duct diameter. Where flow velocities are high (>3m/s), consider using cloth faced absorbent. Duct vibration can usually be treated by damping.

Fan Speed

Typical Applications

Axial or centrifugal flow fans.

Technique

Fan noise is roughly proportional to the 5th power of fan speed. So in many cases it is possible to achieve a large noise reduction from a small drop in fan speed by changing control systems or pulley sizes and re-setting dampers. The following table provides a guide to the trade-off that can be expected.

Fan Speed Reduction	Noise Reduction
10%	2 dB
20%	5 dB
30%	8 dB
40%	11 dB
50%	15 dB

Pneumatic Exhausts

A well-designed silencer will not increase system back pressure.

Almost invariably it is possible to reduce pneumatic exhaust noise permanently by 10 - 30 dB by fitting effective silencers. The following are the practical points that can make the difference between success and failure:

- Back pressure: Fit a larger coupling and silencer.

- Clogging: Fit a straight-through silencer that cannot clog (and has no back pressure).

- Multiple exhausts: Manifold them into a single, larger diameter pipe fitted with the rear silencer from virtually any make of car (from your local tyre and exhaust fitter). Typically 25 dB reduction.

Pneumatic Nozzles

Typical Applications

Cooling, drying and blowing.

Technique

In most cases, it is possible to replace existing nozzles (usually simple copper pipe outlets) for quiet, high efficiency units. These not only reduce noise levels by up to 10 dB, but also use less compressed air. The types of nozzle to look out for are entraining units (schematic below) from various manufacturers and in a variety of sizes.

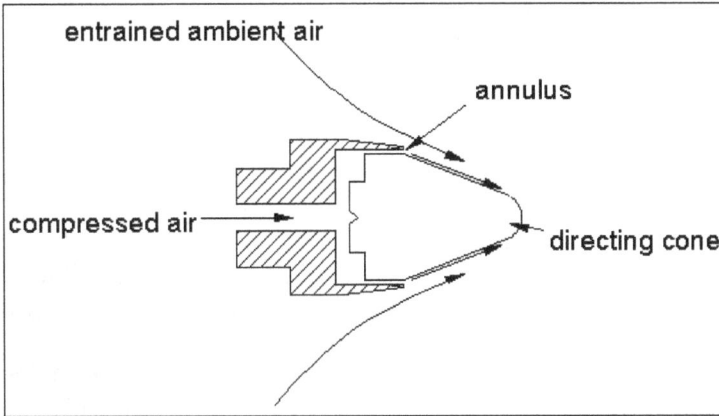

Vibration Isolation Pads

Typical Applications

Machine feet, pumps and mezzanine installations.

Technique

Mounting motors, pumps, gearboxes and other items of plant on rubber bonded cork (or similar) pads can be a very effective way of reducing transmission of vibration and therefore noise radiated by the rest of the structure. This is particularly the case where vibrating units are bolted to steel supports or floors. However, a common error with the use of these pads is for the bolt to "short-circuit" the pad, resulting in no isolation. Additional pads must be fitted under the bolt heads as shown below.

There are many types of off-the-shelf anti-vibration mounts available, for instance rubber/neoprene or spring types. The type of isolator that is most appropriate will depend on, among other factors, the mass of the plant and the frequency of vibration to be isolated. Any supplier of anti-vibration mounts will be able to advise you on this.

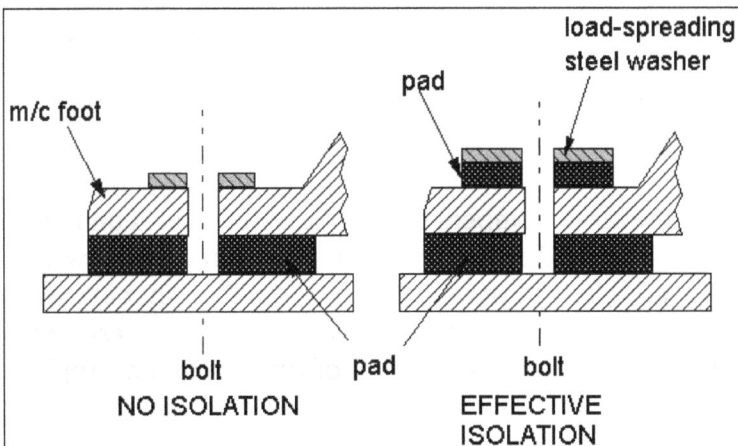

Existing Machine Guards

Technique

The existing guards on many machines can often be improved to provide a significant noise reduction. The two principals involved, which must be used in combination, are:

- Minimise gaps: Reducing by half the "gap" open area in a set of guards can reduce the noise by 3 dB. If you can reduce the openings (flexible seals, additional close fitting panels etc) by 90%, then a 10 dB noise reduction is possible.

- Acoustic absorbent: Lining a significant proportion of the inside of the guards with acoustic absorbent (foam, rockwool/fibreglass) will reduce the noise "trapped" by the guards. Consequently, less noise will escape through any gaps. Failure to line the inside of the guards could result in an increase in noise at the operator's position if the gaps have been minimised.

In most cases, both sets of modifications can be tested in mock-up form using cardboard (and wide tape) to extend the guarding and temporarily fitting areas of acoustic foam inside. Not only does this process help with the practical aspects (access, visibility etc), but it usually also provides a very good indication of the noise reduction that can be expected. Very "Blue Peter" but very effective. Guard vibration radiated as noise can also be treated via damping.

Chain and Timing Belt Drives

Technique

Noisy chain drives can often be replaced directly with quieter timing belts. Within the range of timing belts available, there are also quiet designs that use different tooth profiles to minimise noise. There is also a very new design of belt for applications where noise is critical which uses a chevron tooth pattern to provide very quiet running. Noise reductions in the range of 6 - 20 dB are often possible using this approach.

Electric Motors

Technique

Most companies have large numbers of electric motors used on anything from fans to pumps to machine tools. However, it is not very common knowledge that general duty motors are available (at little or no cost premium) that are up to 10 dB(A) or more quieter than typical units as direct replacements. The best approach is to feed these motors into the system over a period of time so that all replacement motors are quiet motors.

Architectural Acoustics

"Acoustics" in architecture means improving sound in environments. Although it is a complex science, to understand the basics, is to understand that there are two technical categories used in acoustics: soundproofing and acoustical treatment. Soundproofing means "less noise" and treatment, "better sound".

Soundproofing is commonly used in music recording studios - but it can also be applied in locations near major avenues, schools, construction zones or even drummers' neighbors. Soundproofing an environment is like protecting it against bad weather: the structure should be as solid as possible and without holes or cracks. To reduce the noise coming into and going out of a room, one must increase the structural mass of the walls, floor and ceiling, and seal the air gaps surrounding doors and windows, as well as the openings for refrigeration and electrical outlets. The extent of the measures taken will depend on how much noise there is on the outside, and how much you want it to be reduced on the inside.

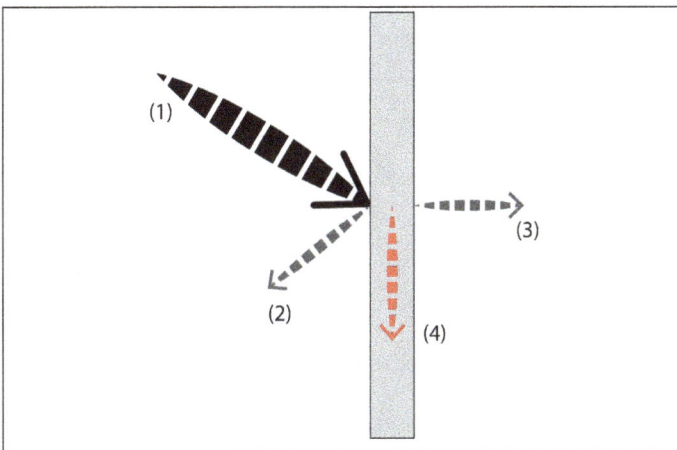

(1) Incident Sound/(2) Reflected Sound/(3) Transmitted Sound/(4) Absorbed Sound.

On the other hand, sound treatments are used when you want to improve sound quality within an environment - for diners to hear and understand conversations at their tables in a restaurant, for students to understand teachers, for the whole audience to enjoy the music in an auditorium. All building materials have acoustic properties as they can potentially absorb, reflect or transmit sounds that reach them. When sounds are reflected, they cause an increase in the overall echo and reverberation levels in a space. When treating rooms correctly, echo and reverberation is reduced - and to treat rooms, there are two methods available: sound absorption and diffusion. The best treatment strategies combine these two techniques.

Sound absorption is defined as the incident sound that strikes a material that is not reflected back. An open window is an excellent absorber, since sounds that pass through the open window are not reflected back. Acoustic absorbers use materials designed for the purpose of absorbing sound that could otherwise be reflected back into the room. The more fibrous a material, the better the absorption, and denser materials are usually less absorbent. The acoustic absorption characteristics of different materials can vary significantly by frequency. In general, low frequency sounds are very difficult to absorb because of their long wavelengths. However, we are generally less sensitive to low frequency sounds, which mean we often do not need to treat a room for low-frequency absorption.

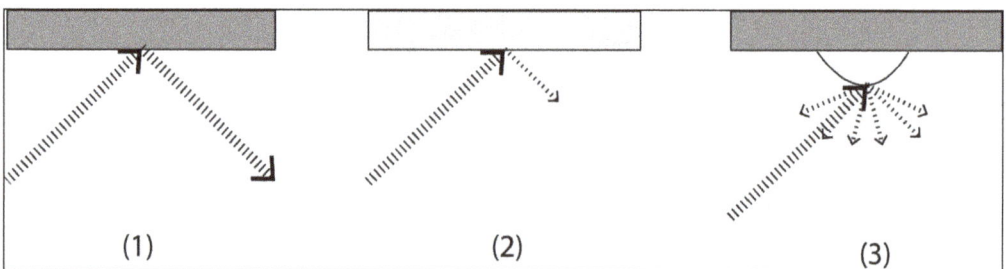

Types of Surfaces: (1) Reflective/(2) Absorptive/(3) Diffusive.

Diffusion is the method of spreading sound energy with a diffuser to improve sound in a space. However, the process and tools of sound diffusion can be misunderstood, even by some professionals. Diffusion spreads the reflected sound energy in a room, also reducing the harmful effects of strong echo and reverberation. One type of diffusor is a curved panel, often with a fabric cover, which can be easily placed on walls and ceilings. These types of panels have the advantage of uniformly spreading flat-wall reflections that would otherwise be combined with original sound waves to create destructive interference. In a concert hall, for example, diffusion panels are used to enhance the richness of sound and help create a sense of spaciousness.

When a sound wave reaches an acoustic material, the sound wave vibrates the fibers or particles of the absorbent material. This vibration causes small amounts of heat due to friction and thus the sound absorption is performed by means of energy for heat conversion. For the majority of conventional acoustic materials, the density and thickness of the material affect the sound absorption amount and frequencies absorbed by the material. While the inherent composition of the acoustic material determines its performance, other factors may be used to improve or influence acoustic performance. For example, the incorporation of an air space behind an acoustic ceiling or a wall panel often improves low frequency performance.

By installing absorbers and diffusers in a space, the level of undesirable noise, in the form of echo and reverberation, is reduced. Noise is a relative term and can range from low levels of intrusive sound in a quiet environment to loud sounds in an already noisy environment. When noise levels are high enough, background noise can mask the speech sound levels that you want to hear. Restaurants can be classic examples of excessive noise interference due to lack of sound absorbing materials to prevent excessive build-up of echo and reverberation. Customers speak louder and louder to be heard and in doing so simply add to the noise. Proper acoustical treatments will help reduce the accumulation of sound reflections and will reduce the need for customers to speak increasingly louder.

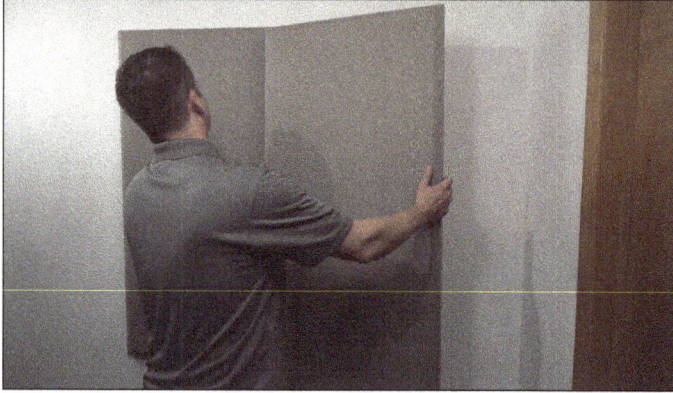

Being in an environment with inadequate acoustics can be extremely unpleasant and directly influences the environmental comfort of a space, our behavior, and even our productivity. Just as architects do not necessarily need be experts in every technical aspect of a project, it is the same for acoustical knowledge – it can be helpful to call on acoustical-product suppliers to carefully review the technical specifications of a project and to recommend the best available products to improve the acoustical environment. However, it's also helpful to have a basic idea of these issues – it will help us make informed decisions for incorporating better sound within the project's design, ultimately delivering a better user experience.

Soundproofing

Soundproofing is the process of developing acoustic treatments to suppress, reflect, diffuse or absorb sound waves either at the source using products such as baffling or insulation on automotive engines, pipes/ducting or machinery rooms, or at the destination using technologies such as double glazed windows and insulation to reduce unwanted external noise from road traffic, airports and other rooms within offices, apartments and other residential dwellings.

Soundproofing Techniques

A range of techniques can be used to reduce sound within a room:

Sound Absorption

Porous or resonant absorbing materials can be installed to convert sound wave energy into a tiny amount of heat effectively reducing the amount of sound that is reflected around a room. Noise can be absorbed by using for sound absorption within restaurants, hotels, schools, home theatres and many other commercial and residential applications.

Sound Diffusion

Irregular surfaces can be used to break down, or scatter sound waves so that they travel along many smaller paths. This result in the overall energy of the sound wave depleting faster and therefore travelling less distance.

Sound Reflection

Reflecting sound waves is a common technique for reducing traffic noise on freeways and other major traffic routes. Typically hard surfaces such as concrete or glass are used to bounce sound waves upwards (into the sky) or away from buildings. The overall architecture of a building including material and angles can also be used to effectively reflect sound.

Soundproofing Categories

Soundproofing falls into two main categories as follows:

- Reducing sound within a room: Soundproof rooms, also known as anechoic (echo free) or semi-anechoic chambers, are designed to reduce unwanted reverberation and echo within a room for the purposes of sound-quality (e.g. recording studios and home theatres) or noise mitigation e.g. hotels, bars, restaurants, and open-plan offices. Room soundproofing typically also includes insulation to reduce unwanted external noise.

- Limiting sound leakage: Soundproofing materials can be utilised to limit the transference of sound waves between rooms (either adjacent, above or below) and from external sources. A range of items within a room or rooms can be soundproofed including windows (double glazing), floors, walls, ceilings and doors.

Absorption (Acoustics)

Sound absorption is the measure of the amount of energy removed from the sound wave as the wave passes through a given thickness of material. Figure is a schematic representation of sound absorption and reflection of an insulating wall. While propagating from air into an absorbing material, the sound wave could experience reflection or absorption thereby losing energy, experiencing dampening effects. In a polymeric material sound absorption takes place by transforming sound waves into heat. Sound absorption is necessary for soundproofing. Materials with their characteristic impedance similar to air are regarded as best soundproofing materials thus foamed plastics are the preferred materials for such applications. Typically, elastomers and amorphous polymers show higher sound absorption properties as compared to semic-rystalline or crystalline materials.

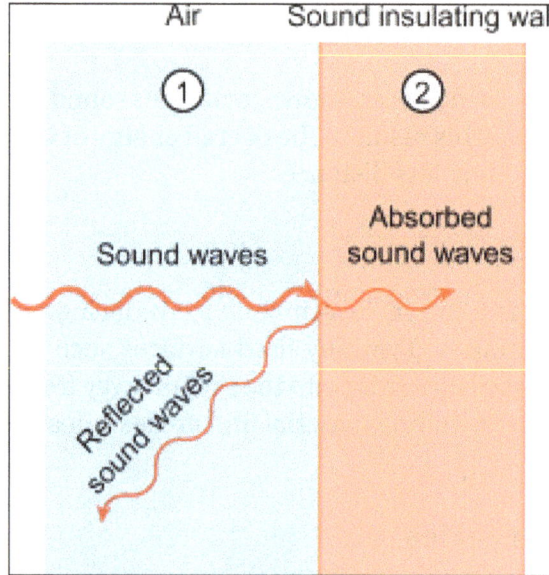

Schematic diagram of sound reflection and
absorption by an insulating wall variables.

Sound absorption materials also need to have great loss of sound energy, air permeability, and refractory quality, as well as the required level of structural strength. Glass wool fibers have poor deformation and decreased sound absorption in the rain, while ceramic materials have poor impact strength. Therefore, metal foams have some obvious advantages for sound absorption, like great stiffness and strength, good refractoriness, good temperature change adaptability, low moisture absorption, and excellent impact energy absorption. Hence, they are widely used for noise and vibration control in aircraft, locomotives, automobiles, machines, and buildings. In buildings, especially office facilities, metal foams can be used for both decoration and sound absorption.

Voice filters are used to reduce or control sound attenuation. These applications range from sound absorption in jet engineering to attenuators in hearing aids. A voice filter with predetermined acoustic resistance can be designed based on the available knowledge. Similar to electrical resistance, sound resistance can be expressed as Resistance = sound pressure/sound speed. The detectable sound pressure is around 2×10^{-5} Pa, and the maximum is 20 Pa. The sound resistance value is related to porosity and pore shape; it increases as the part thickens and decreases as the area grows larger. Metal foam parts are used as acoustic impedance in the telephone transmitter and acceptor to provide the necessary sound resistance.

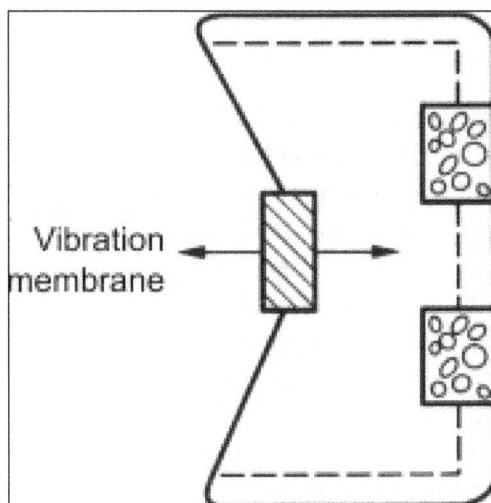

The metal foam sound resistance parts in the telephone transmitter.

The acceleration and weight reduction of the train can cause vibration and noise. Therefore, there are more requirements for noise control in these environments. The application of metal foam can solve these problems in both cars and trains.

A sound wave is one kind of vibration, and it is absorbed by materials with sound absorption through the dispersion and interference of waves in the metal foam. Metal foams are applied in gas or steam tubes for sound absorption. For instance, a large amount of noise is generated during gas transportation in the long-range high-pressure tube, and this can be eliminated by metal foam materials. When applied in power plants of steam turbines and gas-powered machines, Cu foams with relative density of 5% can act as a muffler to eliminate the noise and also ensure the air flow flux.

Powder metallurgical porous metals are commonly used for noise reduction, damping of pressure pulse, and control of mechanical vibration; i.e., control of abrupt pressure changes in compressors and gas-powered machines. All of the open-cell porous materials have the capability of selective damping based on the frequencies of sound waves. The metal foams prepared by investment casting and deposition processes are more efficient and are developed to replace conventional parts made of powder metallurgical porous metals.

Sound waves can be affected and their transmission paths can be changed with the application of open-cell metal foams in the shape of lenses or cylinders. These open-cell metal foams can act as receptors during ultrasound inspection due to the appropriate range of ultrasound impedance, while closed-cell foams can be used to pick up impedance for the ultrasound source.

For different kinds of sound absorption materials, the fibers have good sound absorption, but they have poorer physical properties than metal foams. The wood fiber panels and the micro-perforated panels with good noise reduction capability have strength and stiffness limitations. Metal foams have good properties in comprehensive aspects and a wide range of applications for damping and noise reduction in automobiles, shipping, and aircraft.

For some metals with greater mechanical strength and thermal stability, the foams made of these metals have not only good sound absorption, but also good mechanical properties and thermal conductivity. The sound absorption property for the metal foams is comparable to polymer foams, and it can be maintained at elevated temperatures; consequently, these foams can be employed under more severe environments. For instance, Cu foams can work at temperatures of above 900 °C and the W and Cr foams can be used at even higher temperatures. The requirements for exhaust noise elimination equipment in gas turbines, such as high efficiency, long lifetimes, and light weight, exclude the application of the normal sound-absorption materials. Light metal foams (like porous Ti) can be used for this purpose because of their high-temperature, high-speed erosion resistance, and corrosion resistance properties.

Traditional porous sound absorption materials such as mineral wool, fiberglass, and perforated panels are generally used in the medium of air, but not in water with pressure and temperature changes. Moreover, the impedance incompatibility with water for these materials is much higher. Rubber materials have the perfect impedance compatibility with water, so they can be used as sound-absorption materials in those environments. However, the deformation of rubber in water may change the absorbed sound frequencies. Metal foams have the features of light weight, great strength, and good impedance comparable to that of water, and they can absorb low-frequency noise effectively in water if filled with the proper viscous fluid. It can be seen that metal foams have more advantages in water in terms of impedance compatibility, water pressure, and temperature change, and no pollution is caused by their use.

References

- Report by the International Institute of Noise Control Engneering Working Party on the effect of regulations on road vehicle noise. Noise/News International 3(1995)6, 85–113

- "Preference for nature in urbanized societies: Stress, restoration, and the pursuit of sustainability," Journal of Social Issues, 63 (1): 79-96

- Technical Assessment of the Effectiveness of Noise Walls. Final report. International INCE Publication 99-1. Noise/News International 7(1999)3, 137–161

- "Sound propagation in street canyons: Comparison between diffusely and geometrically reflecting boundaries." Journal of the Acoustical Society of America. 107 ():1394-1404

- "Effects of road traffic noise and the benefit of access to quietness." Journal of Sound and Vibration 295():40-59

Chapter 6

Sound Level Measurement

Acoustic instruments are used in measuring and assessing the sound levels by the use of sound pressure. Sound level meter or decibel meter is one such instrument used for sound level measurement. The topics elaborated in this chapter will help in gaining a better perspective about the concepts of sound intensity and noise calculation as well as instruments associated with it.

The details of noise measurements must be planned to meet some relevant objective or purpose. Some typical objectives would include:

- Investigating complaints.
- Assessing the number of persons exposed.
- Compliance with regulations.
- Land use planning and environmental impact assessments.
- Evaluation of remedial measures.
- Calibration and validation of predictions.
- Research surveys.
- Trend monitoring.

The sampling procedure, measurement location, type of measurements and the choice of equipment should be in accord with the objective of the measurements.

Instrumentation

The most critical component of a sound pressure meter is the microphone, because it is difficult to produce microphones with the same precision as the other, electronic components of a pressure meter. In contrast, it is usually not difficult to produce the electronic components of a microphone with the desired sensitivity and frequency-response characteristics. Lower quality microphones will usually be less sensitive and so cannot measure very low sound pressure levels. They may also not be able to accurately measure very high sound pressure levels found closer to loud noise sources. Lower quality microphones will also have less well-defined frequency response characteristics. Such lower quality microphones may be acceptable for survey type measurements of overall A-weighted levels, but would not be preferred for more precise measurements, including detailed frequency analysis of the sounds.

Sound pressure meters will usually include both A- and C-weighting frequency-response curves. The uses of these frequency weightings were discussed above. They may also include a linear weighting. Linear weightings are not defined in standards and may in practice be limited by the response of the particular microphone being used. Instead of, or in addition to, frequency response weightings, more complex sound pressure meters can also include sets of standard band pass filters, to permit frequency analysis of sounds. For acoustical measurements, octave and one-third octave bandwidth filters are widely used with centre frequencies defined in standards.

The instantaneous sound pressures are integrated with some time constant to provide sound pressure levels. As mentioned above most meters will include both Fast- and Slow-response times. Fast-response corresponds to a time constant of 0.125 s and is intended to approximate the time constant of the human hearing system. Slow-response corresponds to a time constant of 1 s and is an old concept intended to make it easier to obtain an approximate average value of fluctuating levels from simple meter readings.

Standards classify sound pressure meters as type 1 or type 2. Type 2 meters are adequate for broad band A-weighted level measurements, where extreme precision is not required and where very low sound pressure levels are not to be measured. Type 1 meters are usually much more expensive and should be used where more precise results are needed, or in cases where frequency analysis is required.

Many modern sound pressure meters can integrate sound pressure levels over some specified time period, or may include very sophisticated digital processing capabilities. Integrating meters make it possible to directly obtain accurate measures of LAeq,T values over a user-specified time interval, T. By including small computers in some sound pressure meters, quite complex calculations can be performed on the measured levels and many such results can be stored for later read out. For example, some meters can determine the statistical distribution of sound pressure levels over some period, in addition to the simple LAeq,T value. Recently, hand-held meters that perform loudness calculations in real time have become available. Continuing rapid developments in instrumentation capabilities are to be expected.

Measurement Locations

Where local regulations do not specify otherwise, measurements of environmental noise are usually best made close to the point of reception of the noise. For example, if there is concern about residents exposed to road traffic noise it is better to measure close to the location of the residents, rather than close to the road. If environmental noises are measured close to the source, one must then estimate the effect of sound propagation to the point of reception. Sound propagation can be quite complicated and estimates of sound pressure levels at some distance from the source will inevitably introduce further errors into the measured sound pressure levels. These errors can be avoided by measuring at locations close to the point of reception.

Measurement locations should normally be selected so that there is a clear view of the sound source and so that the propagation of the sound to the microphone is not shielded or blocked by structures that would reduce the incident sound pressure levels. For example, measurements of aircraft noise should be made on the side of the building directly exposed to the noise. The position of the measuring microphone relative to building façades or other sound-reflective surfaces is also important and will significantly influence measured sound pressure levels. If the measuring microphone is located more than several meters from reflecting surfaces, it will provide an unbiased indication of the incident sound pressure level. At the other extreme, when a measuring microphone is mounted on a sound-reflecting surface, such as a building façade, sound pressure levels will be increased by 6 dB, because the direct and reflected sound will coincide. Some standards recommend a position 2 m from the façade and an associated 3 dB correction. The effect of façade reflections must be accounted for to represent the true level of the incident sound. Thus, while locating the measuring microphone close to the point of reception is desirable, it leads to some other issues that must be considered to accurately interpret measurement results. Where exposures are measured indoors, it is necessary to measure at several positions to characterize the average sound pressure level in a room. In other situations, it may be necessary to measure at the position of the exposed person.

Sampling

Many environmental noises vary over time, such as for different times of day or from season to season. For example, road traffic noise may be considerably louder during some hours of the day but much quieter at night. Aircraft noise may vary with the season due to different numbers of aircraft operations. Although permanent noise monitoring systems are becoming common around large airports, it is usually not possible to measure sound pressure levels continuously over a long enough period of time to completely define the environmental noise exposure. In practice, measurements usually only sample some part of the total exposure. Such sampling will introduce uncertainties in the estimates of the total noise exposure.

Traffic noise studies have identified various sampling schemes that can introduce errors of 2-3 dB in estimates of daytime LAeq,T values and even larger errors in nighttime sound pressure levels. These errors relate to the statistical distributions of sound pressure levels over time. Thus, the sampling errors associated with road traffic noise may be quite different from those associated with other noise, because of the quite different variations of sound pressure levels over time. It is also difficult to give general estimates of sampling errors due to seasonal variations. When making environmental noise measurements it is important that the measurement sample is representative of all of the variations in the noise in question, including variations of the source and variations in sound propagation, such as due to varying atmospheric conditions.

Calibration and Quality Assurance

Sound pressure meters can be calibrated using small calibrated sound sources. These devices are placed on the measurement microphone and produce a known sound pressure level with a specified accuracy. Such calibrations should be made at least daily, and more often if there is some possibility that handling of the sound pressure meter may have modified its sensitivity. It is also important to have a complete quality assurance plan. This should require annual calibration of all noise measuring equipment to traceable standards and should clearly specify correct measurement and operating procedures.

Source Characteristics and Sound Propagation

To make a correct assessment of noise it is important to have some appreciation of the characteristics of environmental noise sources and of how sound propagates from them. One should consider the directionality of noise sources, the variability with time and the frequency content. If these are in some way unusual, the noise may be more disturbing than expected. The most common types of environmental noise sources are directional and include: road-traffic noise, aircraft noise, train noise, industrial noise and outdoor entertainment facilities. All of these types of environmental noise are produced by multiple sources, which in many cases are moving. Thus, the characteristics of individual sources, as well as the characteristics of the combined sources, must be considered.

For example, we can consider the radiation of sound from individual vehicles, as well as from a line of vehicles on a particular road. Sound from an ideal point source (i.e. non-directional source) will spread out spherically and sound pressure levels would decrease 6 dB for each doubling of distance from the source. However, for a line of such sources, or for an integration over the complete pass-by of an individual moving source, the combined effect leads to sound that spreads cylindrically and to sound pressure levels that decrease at 3 dB per doubling of distance. Thus, there are distinct differences between the propagation of sound from an ideal point source and from moving sources. In practice one cannot adequately assess the noise from a fixed source with measurements at a single location; it is essential to measure in a number of directions from the source. If the single source is moving, it is necessary to measure over a complete pass-by, to account for sound variation with direction and time.

In most real situations this simple behaviour is considerably modified by reflections from the ground and from other nearby surfaces. One expects that when sound propagates over lose ground, such as grass, that some sound energy will be absorbed and sound pressure levels will actually decrease more rapidly with distance from the source.

Although this is approximately true, the propagation of sound between sources and receivers close to the ground is much more complicated than this. The combination of direct and ground-reflected sound can combine in a complex manner which can lead to strong cancellations at some frequencies and not at others. Even at quite short source-to-receiver distances, these complex interference effects can significantly modify the propagating sound. At larger distances (approximately 100 m or more), the propagation of sound will also be significantly affected by various atmospheric conditions. Temperature and wind gradients as well as atmospheric turbulence can have large effects on more distant sound pressure levels. Temperature and wind gradients can cause propagating sound to curve either upwards or downwards, creating either area of increased or decreased sound pressure levels at points quite distant from the source. Atmospheric turbulence can randomize sound so that the interference effects resulting from combinations of sound paths are reduced. Higher frequency sound is absorbed by air depending on the exact temperature and relative humidity of the air. Because there are many complex effects, it is not usually possible to accurately predict sound pressure levels at large distances from a source.

Using barriers or screens to block the direct path from the source to the receiver can reduce the propagation of sound. The attenuating effects of the screen are limited by sound energy that diffracts or bends around the screen. Screens are more effective at higher frequencies and when placed either close to the sound source or the receiver; they are less effective when placed far from the receiver. Although higher screens are better, in practice it is difficult to achieve more than about a 10 dB reduction. There should be no gaps in the screen and it must have an adequate mass per unit area. A long building can be an effective screen, but gaps between buildings will reduce the sound attenuation.

In some cases, it may be desirable to estimate environmental sound pressure levels using mathematical models implemented as computer programmes. Such computer programmes must first model the characteristics of the source and then estimate the propagation of the sound from the source to some receiver point. Although such prediction schemes have several advantages, there will be some uncertainty as to the accuracy of the predicted sound pressure levels. Such models are particularly useful for road traffic noise and aircraft noise, because it is possible to create data bases of information describing particular sources. For more varied types of noise, such as industrial noise, it would be necessary to first characterize the noise sources. The models then sum up the effects of multiple sources and calculate how the sound will propagate to receiver points. Techniques for estimating sound propagation are improving and the accuracy of these models is also expected to improve. These models can be particularly useful for estimating the combined effect of a large number of sources over an extended period of time. For example, aircraft noise prediction models are typically used to predict average yearly noise exposures, based on the combination of aircraft events over a complete year. Such models can be applied to predict sound pressure level contours around airports for these average yearly conditions. This is of course much less expensive than measuring at many locations over a complete one year-period. However, such models can be quite complex, and require skilled users and accurate data bases. Because environmental noise prediction models are still developing, it is advisable to confirm predictions with measurements.

Sound Transmission into and within Buildings

Sources of environmental noise are usually located outdoors; for example, road traffic, aircraft or trains. However, people exposed to these noises are often indoors, inside their home or some other building. It is, therefore, important to understand how environmental noises are transmitted into buildings. Most of the same fundamentals discussed earlier apply to airborne sound propagation between homes in multifamily dwellings, via common walls and floors. However, within buildings we can also consider impact sound sources, such as footsteps, as well as airborne sounds.

The amount of incident sound that is transmitted through a building façade is measured in terms of the sound reduction index. The sound reduction index, or transmission loss, is defined as 10 times the logarithm of the ratio of incident-to-transmitted sound power, and it describes in decibels how much the incident sound is reduced on passing through a particular panel. This index of constructions usually increases with the frequency of the incident sound and with the mass of the construction. Thus, heavier or more massive constructions tend to have higher sound reductions. When it is not possible to achieve the desired transmission loss by increasing the mass of a panel, increased sound reduction can be achieved by a double panel construction. The two layers should be isolated with respect to vibrations and there should be sound absorbing material in the cavity. Such double panel constructions can provide much greater sound reduction than a single panel. Because sound reduction is also greater at higher frequencies most problems occur at lower frequencies, where most environmental noise sources produce relatively high sound pressure levels.

The sound reduction of buildings can be measured in standard laboratory tests, where the test panel is constructed in an opening between two reverberant test chambers. In these tests sound fields are quite diffuse in both test chambers and the sound reduction index is calculated as the difference between the average sound pressure levels in the two rooms, plus a correction involving the area of the test panel and the total sound absorption in the receiving room. The sound reduction of a complete building façade can also be measured in the field using either natural environmental noises or test signals from loudspeakers. In either case the noise, as transmitted through the façade, must be greater in level than other sounds in the receiving room. For this outdoor-to-indoor sound propagation case, the measured sound reduction index will also depend on the angle of incidence of the outdoor sound, as well as the position of the outdoor measuring microphone relative to the building façade. Corrections of up to 6 dB must be made to the sound pressure level measured outdoors, to account for the effect of reflections from the façade.

The sound reduction of most real building façades is determined by a combination of several different elements. For example, a wall might include windows, doors or some other type of element. If the sound reduction index values of each element are known, the values for the combined construction can be calculated from the area-weighted sums of the sound energy transmitted through each separate element. Although parts of the building façade, such as massive wall constructions, can be very effective barriers to sound, the sound reduction index of the complete façade is often greatly reduced by less effective elements such as windows, doors or ventilation openings. Completely open windows as such would have a sound reduction index of 0 dB. If window openings make up 10% of the area of a wall, the sound reduction index of the combined wall and open window could not exceed 10 dB. Thus it is not enough to specify effective sound reducing façade constructions, without also solving the problem of adequate ventilation that does not compromise the sound transmission reduction by the building façade.

Sound reduction index values are measured at different frequencies and from these, single number ratings are determined. Most common are the ISO weighted sound reduction index and the equivalent ASTM sound transmission class. However, in their original form these single number ratings are only appropriate for typical indoor noises that usually do not have strong low frequency components. Thus, they are usually not appropriate single number ratings of the ability of a building façade to block typical environmental noises. More recent additions to the ISO procedure have included source spectrum corrections intended to correct approximately for other types of sources. Alternatively, the ASTM-Outdoor-Indoor Transmission Class rating calculates the A-weighted level reduction to a standard environmental noise source spectrum. Within buildings the impact sound insulation index can be measured with a standard impact source and determined according to ISO and ASTM standards.

More Specialized Noise Measures

Loudness and Perceived Noise Levels

There are procedures to accurately rate the loudness of complex sounds. These usually start from a 1/3 octave spectrum of the noise. The combination of the loudness contributions of each 1/3 octave band with estimates of mutual masking effects, leads to a single overall loudness rating in sones. A similar system for rating the noisiness of sounds has also been developed. Again a 1/3 octave spectrum of the noise is required and the 1/3 octave noise levels are compared with a set of equal-noisiness contours. The individual 1/3 octave band noisiness estimates are combined to give an overall perceived noise level (PNL) that is intended to accurately estimate subjective evaluations of the same sound. The PNL metric was initially developed to rate jet aircraft noise.

PNL values will vary with time, for example when an aircraft flies by a measuring point. The effective perceived noise level measure (EPNL) is derived from PNL values and is intended to provide a complete rating of an aircraft fly-over. EPNL values add both a duration correction and a tone correction to PNL values. The duration correction ensures that longer duration events are rated as more disturbing. Similarly, noise spectra that seem to have prominent tonal components are rated as more disturbing by the tone-correction procedure. There is some evidence that these tone corrections are not always successful in improving predictions of adverse responses to noise events. EPNL values are used in the certification testing of new aircraft. These more precise measures ensure that the noise from new aircraft is rated as accurately as possible.

Aviation Noise Measures

There are many measures for evaluating the long-term average sound pressure levels from aircraft near airports. They include different frequency weightings, different summations of levels and numbers of events, as well as different time-of-day weightings. Most measures are based on either A-weighted or PNL-weighted sound pressure levels. Because of the many other large uncertainties in predicting community response to aircraft noise, there seems little justification for using the more complex PNL-weighted sound pressure levels and there is a trend to change to A-weighted measures.

Most aviation noise measures are based on an equal energy approach and hence they sum up the total energy of a number of aircraft fly-overs. However, some older measures were based on different combinations of the level of each event and the number of events. These types of measures are gradually being replaced by measures based on the equal energy hypothesis such as LAeq,T values. There is also a range of time-of-day weightings incorporated into current aircraft noise measures. Night-time weightings of 6–12 dB are currently in use. Some countries also include an intermediate evening weighting.

The day-night sound pressure level L_{dn} is an LAeq,T based measure with a 10 dB night-time weighting. It is based on A-weighted sound pressure levels and the equal energy principle. The noise exposure forecast (NEF) is based on the EPNL values of individual aircraft events and includes a 12 dB night-time weighting. It sums multiple events on an equal energy basis. However, the Australian variation of the NEF measure has a 6 dB evening weighting and a 6 dB night-time weighting. The German airport noise equivalent level (LEQ(FLG)) is based on A-weighted levels, but does not follow the equal energy principle.

The weighted equivalent continuous perceived noise level (WECPNL) measure proposed by ICAO is based on the equal energy principle and maximum PNL values of aircraft fly-overs. However, in Japan an approximation to this measure is used and is based on maximum A-weighted levels. The noise and number index (NNI), formerly used in the United Kingdom, was derived from maximum PNL values but was not based on the equal energy principle. An approximation to the original version of the NNI has been used in Switzerland and is based on maximum A-weighted levels of aircraft fly-overs, but its use will soon be discontinued. Changes in these measures are slow because their use is often specified in national legislation. However, several countries have changed to measures that are based on the equal energy principle and A-weighted sound pressure levels.

Impulsive Noise Measures

Impulsive sounds, such as gun shots, hammer blows, explosions of fireworks or other blasts, are sounds that significantly exceed the background sound pressure level for a very short duration. Typically each impulse lasts less than one second. Measurements with the meter set to 'Fast' response do not accurately represent impulsive sounds. Therefore the meter response time must be shorter to measure such impulse type sounds. C-weighted levels have been found useful for ratings of gun shots. Currently no mathematical description exist which unequivocally defines impulsive sounds, nor is there a universally accepted procedure for rating the additional annoyance of impulsive sounds. Future versions of ISO Standard 1996 are planned to improve this situation.

Measures of Speech Intelligibility

The intelligibility of speech depends primarily on the speech-to-noise ratio. If the level of the speech sounds are 15 dB or more above the level of the ambient noise, the speech intelligibility at 1 m distance will be close to 100%. This can be most simply rated in terms of the speech-to-noise ratio of the A-weighted speech and noise levels. Alternatively, the speech intelligibility index (formerly the articulation index) can be used if octave or 1/3 octave band spectra of the speech and noise are available.

When indoors, speech intelligibility also depends on the acoustical properties of the space. The acoustical properties of spaces have for many years been rated in terms of reverberation times. The reverberation time is approximately the time it takes for a sound in a room to decrease to inaudibility after the source has been stopped. Optimum reverberation times for speech have been specified as a function of the size of the room. In large rooms, such as lecture halls and theaters, a reverberation time for speech of about 1 s is recommended. In smaller rooms such as classrooms, the recommended value for speech is about 0.6 s. More modern measures of room acoustics have been found to be better correlates of speech intelligibility, and some combine an assessment of both the speech/noise ratio and room acoustics. The most widely known is the speech transmission index (STI), or the abbreviated version of this measure referred to as RASTI. In smaller rooms, such as school classrooms, the conventional approach of requiring adequately low ambient noise levels, as well as some optimum reverberation time, is probably adequate to ensure good speech intelligibility. In larger rooms and other more specialized situations, use of the more modern measures may be helpful.

Indoor Noise Ratings

The simplest procedure for rating levels of indoor noise is to measure them in terms of integrated A-weighted sound pressure levels, as measured by LAeq,T. This approach has been criticized as not being the most accurate rating of the negative effects of various types of noises, and is thought to be particularly inadequate when there are strong low-frequency components. Several more complex rating schemes are available based on octave band measurements of indoor noises. In Europe the noise rating system, and in North America the noise criterion, both include sets of equal-disturbance type contours. Measured octave band sound pressure levels are compared with these contours and an overall noise rating is determined. More recently, two new schemes have been proposed: the balanced noise criterion procedure and the room criterion system. These schemes are based on a wider range of octave bands extending from 16–8 000 Hz. They provide both a numerical and a letter rating of the noise. The numerical part indicates the level of the central frequencies important for speech communication and the letter indicates whether the quality of the sound is predominantly low-, medium- or high-frequency in nature. Extensive comparisons of these room noise rating procedures have yet to be performed. Because the newer measures include a wider range of frequencies, they can better assess a wider range of noise problems.

Sound-level Meter

Sound-level meter is a device for measuring the intensity of noise, music, and other sounds. A typical meter consists of a microphone for picking up the sound and converting it into an electrical signal, followed by electronic circuitry for operating on this signal so that the desired characteristics can be measured. The indicating device is usually a meter calibrated to read the sound level in decibels (dB; a logarithmic unit used to measure the sound intensity). Threshold of hearing is about zero decibels for the average young listener, and threshold of pain (extremely loud sounds) is around 120 decibels, representing a power 1,000,000,000,000 (or 10^{12}) times greater than zero decibels.

The electronic circuitry can be adjusted to read the level of most frequencies in the sound being measured or the intensity of selected bands of frequencies. Because the alternating current (AC) signal received by the unit's microphone first must be converted to a direct current (DC), a time constant must be incorporated to average the signal. The constant selected depends on the purpose for which the instrument was designed or for which it is being used.

A typical sound-level meter can be switched between a scale that reads sound intensities uniformly for most frequencies—called unweighted—and a scale that introduces a frequency-dependent weighting factor, thus yielding a response more nearly like that of the human ear. A-frequency-weighting is the most commonly used standard, but B-, C-, D-, and Z-frequency-weightings also exist. The A-frequency-weighting scale is useful in describing how complex noises affect people. Thus, the scale is recognized internationally for measurements relating to prevention of deafness from excessive noise in work environments.

In the early 1970s, as concern about noise pollution increased, accurate, versatile, portable noise-measuring instruments were developed. Sound level is not a measure of loudness, as loudness is a subjective factor and depends on the characteristics of the ear of the listener. In an attempt to overcome this problem, scales have been developed to correlate loudness with objective measurements of sound. The Fletcher–Munson curve, for example, shows the relationship between loudness in decibels and subjectively judged loudness. Other variables have also been studied.

Noise Calculation

Sound power is the acoustical power, sometimes expressed in watts, generated by a source of sound such as a hydraulic pump. Two sound sources (pumps) could be compared by their relative acoustical wattage outputs, but can be more conveniently compared by assigning decibel ratings on the dB power scale.

Sound pressure is the strength of the traveling sound wave, in PSI or other pressure units, at a specified distance from the source of the sound. Discomfort or damage to a listener's ears is from sound pressure, not from sound power at the source. Although two sound waves could be compared by their PSI as read on a sound level meter, it is more convenient to compare them on the dB pressure scale. This is a different scale than the dB power scale on which acoustical wattage is rated.

The dB scales for expressing and comparing sound power and sound pressure were arbitrarily selected so as to be convenient to use. The two scales are different but were carefully defined so their relationship to each other would be such that a change of (so many) dB on the power level at the source would result in the same dB change in pressure reading at any distance from the source. Instead of a linear scale, dB ratings were placed on a logarithmic scale to compress the upper end into a more usable and practical range.

A dB has no assigned value; it does not represent a definite quantity of anything; it is simply a ratio used to compare the relative intensities of two sound waves, or the relative power levels of two sound sources. It can be, and is, used in other technologies such as electronics for comparing voltage, current, or power levels. Decibels could be used in illumination for comparing power level of two sources of light, or the illumination intensity at distances from the light source. They could be used in mechanical work for comparing two levels of force, torque, work, HP, etc.

Whether or not so stated, a dB rating is always a ratio either between two sounds, or between one sound and a fixed reference base. To state that a certain pump has an 80 dB noise level means that the ratio between its sound level and that of the selected zero base (0 dB) is 80 on the dB scale.

The dB Pressure Scale

The standard reference base selected for the intensity (pressure) of a sound wave is 0.000,000,003 PSI because this is the least sound wave intensity that can be detected by the average human ear. It is also the least difference between two sound waves that can normally be detected. This pressure has been assigned a value of 0 dB. It is more pressure has been assigned a value of 0 dB. It is more conveniently written in exponent form as 3×10^{-9} PSI. All other sounds are louder than this, the loudest being over 10,000,000 times the PSI intensity of the 0 dB base. Because this wide range of sound pressure is unhandy to work with, the dB scale, instead of being linear, was set up in logarithms to compress it into a more compact scale where not more than 3 digits are required to express pressure levels over the entire audible range. When comparing the sound pressure of two sound waves, their dB difference is, by definition, 20 times the logarithm (to the base 10) of their ratio.

The dB Power Scale

Sound radiation power is also compared by dB ratings, and for this purpose the 0 dB reference base was arbitrarily chosen to be 0.000,000,000,001 watt, or written in exponent form: 1×10^{-12} watt. Since sound pressure decreases as the square of the distance to the source, the power scale was set up as the square root of the pressure scale, so that any change in dB power at the source would cause the same dB pressure change at any distance. So, the dB power scale was defined as 10 (instead of 20) times the logarithm (to the base 10) of the ratio between two acoustic power levels. (To take the square root of a number, its logarithm is divided by 2).

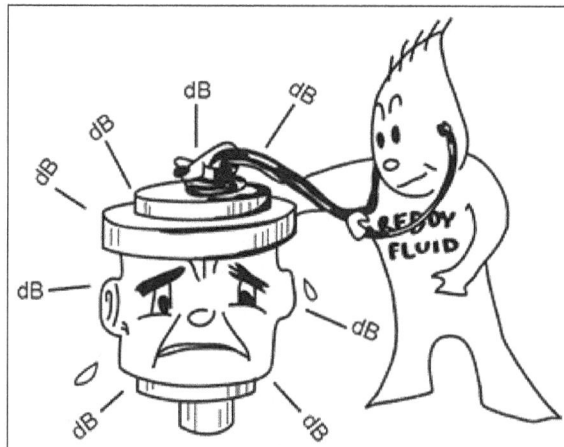

This issue shows how to make calculations relating to noise level of one or several hydraulic pumps.

1. Comparing Two Pumps for Noise: We can compare the dB rating of one pump against another and select the one with less noise, but this does not really give us a "feel" for how noisy each one would be. Standardization methods are under development for specifying noise in "sones". If adopted by all manufacturers it would be easy to compare one pump with another because the assigned value in sones is the loudness as it appears to a listener. It is derived experimentally from data averaged from many listeners. Two sones sound twice as loud as one sone, three sones sound three times as loud as one, etc.

To find the actual acoustic power difference between two pumps, their manufacturer's dB ratings can be converted into watts.

2. Calculation of Acoustical Power of a Pump: Acoustical power, either in dB or in watts, cannot be measured by any known method; it can only be calculated by taking a dB sound pressure reading with a sound level meter at a known distance from the center of the pump, then using the formula:

$$\text{dB power} = \text{dB pressure} + 20 \log \text{distance (feet)} - 2.5 \text{ dB}$$

The factor -2.5 dB is an approximation to take care of moderate sound reflections from walls.

Example: Find the dB noise power of a pump from a meter reading of 87 dB pressure taken 9 feet from the pump.

Solution: dB power = 87 + [20 × 0.954] - 2.5 = 103.58 dB.

The acoustical wattage of the pump can be calculated from Paragraph 8 after its dB power level is found.

3. Calculation of dB Pressure: If the dB rating of a pump is known, the sound pressure it will produce at any distance from its center can be calculated as follows:

$$\text{dB pressure} = \text{dB power} - 20 \log \text{distance (feet)} + 2.5 \text{ dB}.$$

Example: A pump rated at 87 dB (power) will give a dB meter reading at 12 feet: dB (pressure) = 87 - [20 x 1.079] + 2.5 = 67.9 dB.

4. Decrease of Sound Pressure with Distance:

Example: Find the decrease in dB pressure if the distance to the pump is increased to 3 times the original distance.

Solution: dB decrease = 20 log 3 = 9.54 dB. This means that the dB pressure level has dropped 9.54 dB from what it was at the original distance.

5. Increase in Noise Caused by a Second Source: If a hydraulic power unit having several pumps is proposed, what will be the combined noise level with all pumps running?

The dB rating of each pump must be obtained from its manufacturer. These dB ratings, being logarithmic, cannot be added directly. The easiest way to combine them is with the chart in the next column. After the total radiated power has been determined, then follow the method of Paragraph 3 to find the pressure level produced at a specified distance.

Because of the way dB pressure and dB power are defined, any change in dB power at the source (such as adding additional pumps) will cause the same change in dB pressure level at any distance from the pump location.

If a second pump or power unit is to be installed, either at the same or a different location than the existing power unit, its effect on the dB pressure level at a listening location where an operator is working can be determined by the method.

Difference in dB Levels									
0	1	2	3	4	5	6	7	8	10
dB to be Added to Higher Level									
3.0	2.6	2.1	1.8	1.5	1.2	1.0	0.8	0.6	0.4

Example: If noise from Pump 1 is 85 dB, and Pump 2 is added with 90 dB, what will be the new noise level?

Solution: 90 - 85 = 5 dB difference. Use the chart: Add 1.2 dB + 90 dB = 91.2 dB.

In combining several noise sources, combine the two highest first. Combine this total with the next highest level, etc.

6. Pump Added to a Noisy Background: To find the additional noise which will be contributed by a pump added to an already noisy background, first measure the existing noise level at the listening location. Then, using the manufacturer's dB rating for the pump to be added, calculate what its dB pressure level would be at the listening location if run by itself. Finally, combine the two noise level by using the chart above.

7. Finding the dB Power Level: Use this method to find the dB power level when the output power, in watts, is known. This conversion would be used more often in electronics than in acoustics.

Example: Assume a 15-watt power level. What is this in dB? The base for 0 dB is taken as 1×10^{-12} watts.

Solution: First, find the ratio of 15 watts to the 0 dB base. This can be calculated as follows: $15 \div 10^{-12} = 1.5 \times 10^{13}$. Then, take log of $1.5 \times 10^{13} = 13.176$. Finally, multiply by 10 = 131.76 dB. This means that an amplifier delivering 15 watts is working at 131.76 dB above 0 dB.

8. Finding Wattage of a Certain dB Level: This, too, may be used more in electronics. If the dB power level above 0 dB is known, the wattage can be calculated:

- Assume known dB level is 125. Divide by 10 = 12.5 dB.

- Find anti-log of 12.5 = 3.16×10^{12}.

- Multiply by wattage of 0 dB: $3.16 \times 10^{12} \times 10^{-12} = 3.16w$.

9. Finding dB Gain: This is used in electronics to find the power gain of an amplifier. The input and output levels, in watts, must be known. Then proceed as follows:

- Assume an input power of 1 milliwatt (0.001 watt) and an output power of 50 watts. Find the ratio: $50 \div 0.001 = 50,000$. Then, take the log of 50,000 = 4.699. Finally, multiply by 10 = 46.99 dB power gain.

- To find voltage gain, input and output voltages (converted to equivalent voltages across identical input and output resistances) must be known. Then, proceed as above except in the final step, multiply by 20 instead of 10. The answer will be in dB voltage gain.

Sound Exposure Levels

Sound Exposure (E) is the (sound-pressure)^2 measured over a stated period of time or noise event. Measurements are normally A-weighted to relate to the human response.

The SI unit of sound exposure is the $Pa^2 \cdot s$ (pascal-squared second), however Occupational Noise Exposure Meters, also known as noise dose meters, record the $Pa^2 \cdot h$ (pascal-squared hour) levels throughout over the working day.

For a given period of time, an increase of 10 dB(A) in sound pressure level corresponds to a tenfold increase in the sound exposure.

$1\ Pa^2 \cdot h$ = 100% Dose = 85 LAeq(8 h) = 85 dBA for 8 hours.

$0.1\ Pa^2$ = 74.9 LAeq,8 h
$0.2\ Pa^2$ = 78.0 LAeq,8 h
$0.5\ Pa^2$ = 81.9 LAeq,8 h
$1\ Pa^2$ = 84.9 LAeq,8 h = 100% Dose
$2\ Pa^2$ = 88.0 LAeq,8 h
$5\ Pa^2$ = 91.9 LAeq,8 h
$10\ Pa^2$ = 94.8 LAeq,8 h

Sound Exposure Action Values (EAV) is the 8-hour daily exposure to noise above which employers are required to take action to control exposure. For noise there are two action levels,

The Lower EAV is 80 dBA and a peak sound pressure of 135 dBC ≈ LCpeak.
The Upper EAV is 85 dBA and a peak sound pressure of 137 dBC ≈ LCpeak.

Sound Exposure Limit Value (ELV) is the maximum noise an employee may be exposed to on any single 8-hour day is 87 dBA and a peak sound pressure of 140 dBC.

Sound Exposure Definition IEC 801-21-23, time integral of the squared A-weighted, instantaneous sound pressure, over a stated period of time or event. The frequency weighting may be other than A, if so specified.

- Duration of integration is implicitly included in the integral, and need not be reported explicitly.

- The unit of sound exposure is the pascal-squared-second ($Pa^2 \cdot s$), if time is in seconds; the pascal-squared kilosecond ($Pa^2 \cdot ks$), if time is in kiloseconds; the pascal-squared hour ($Pa^2 \cdot h$), if time is in hours.

Most modern sound level meters measure sound exposure and then calculate the required parameters.

Sound Exposure Level (LE) is the constant sound level that has the same amount of energy in one second as the original noise event. A-weighted sound exposure levels are denoted by the symbol LAE.

Sound Exposure Level is similar to the Leq - equivalent continuous sound level786415 as the total sound energy is integrated over the measurement period. However instead of averaging over the measurement period, reference duration of 1 second is used.

It follows that the sound exposure level = Leq + 10·Log10T where T is in seconds for the whole measurement period.

- Log10 is also written as lg = logarithm to the base 10; as recommended by ISO the International Standards Organisation.

Sound Exposure Level (SEL) is numerically equivalent to the total sound energy. For example a noise level of 90 dBA lasting 1 second would have a SEL of 90 dBA but if the event lasted 2 seconds the SEL would be 93 dBA. Put another way if a second event of 80 dBA occurred it would have to last 10 seconds to register a 90 dBA SEL.

Sound Exposure Levels normalized to 1 second are a very useful way of comparing different sound events and sources.

Sound Exposure Level Definition IEC 801-22-17, logarithm of the ratio of a given time integral of squared A-frequency weighted sound pressure, over a stated time interval or event such as an aircraft flyover, to the product of the squared reference sound pressure of 20 µPa and the reference duration of one second. Sound exposure level in decibels is ten times the logarithm to the base ten of that ratio.

- The reference sound pressure and the frequency weighting may be different, if specifically stated with the sound exposure level.

Sound Exposure Meter a small instrument designed to be worn by an individual to provide a measure of the accumulated sound exposure received by the wearer while moving about during the workday.

The instrument is calibrated in Pa²·h. If the meter is worn for only a representative part of the working day, the reading must be corrected appropriately.

Sound Exposure Meters measuring Pa²·h directly are also known as noise dosimeters.

Noise Dosimeter or Noise Dosemeter is a more general term for instruments having a similar purpose but may be calibrated to suit differing standards around the world.

Sound Intensity

Any piece of machinery that vibrates radiates acoustical energy. Sound power is the rate at which energy is radiated [energy per unit time). Sound intensity describes the rate of energy flow through a unit area. In the SI system of units the unit area is 1 m². And hence the units for sound intensity are Watts per square metre.

Sound intensity also gives a measure of direction as there will be energy flow in some directions but not in others. Therefore sound intensity is a vector quantity as it has both magnitude and direction. On the other hand pressure is a scalar quantity as it has magnitude only. Usually we measure the intensity in a direction normal (at 90°) to a specified unit area through which the sound energy is flowing.

We also need to state that sound intensity is the time-averaged rate of energy flow per unit area. In some cases energy may be travelling back and forth. This will not be measured; if there is no net energy flow there will be no net intensity.

In the diagram opposite the sound source is radiating energy. All this energy must pass through an area enclosing the source. Since intensity is the power per area, we can easily measure the normal spatial-averaged intensity over an area which encloses the source and then multiply it by the area to find the sound power. Note that intensity (and pressure) follows the inverse square law for free field propagation. This can be seen in the diagram; at a distance 2r from the source the area enclosing the source is 4 times as large as the area at a distance r. Yet the power radiated must be the same whatever the distance and consequently the intensity, the power per area, must decrease.

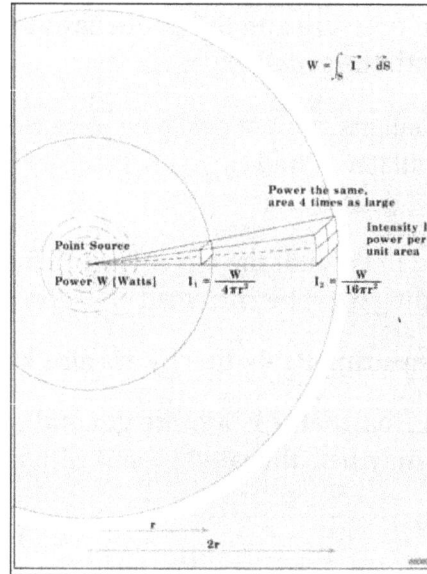

On the factory floor we can make sound pressure measurements and find out if the workers risk hearing damage. But once we have found this, we may well want to reduce the noise. To do this, we need to know how much noise is being radiated and by what machine. We therefore need to know the sound power of the individual machines and rank them in order of highest sound power. Once we have located the machine making most noise we may want to reduce the noise by locating the individual components radiating noise.

We can do all this with intensity measurements. Previously we could only measure pressure which is dependent on the sound field. Sound power can be related to sound pressure only under carefully controlled conditions where special assumptions are made about the sound field. Specially constructed rooms such as anechoic or reverberant chambers fulfil these requirements. Traditionally, to measure sound power, the noise source had to be placed in these rooms.

Sound intensity, however, can be measured in any sound field. No assumptions need to be made. This property allows all the measurements to be done directly in situ. And measurements on individual machines or individual components can be made even when all the others are radiating noise, because steady background noise makes no contribution to the sound power determined when measuring intensity.

Because sound intensity gives a measure of direction as well as magnitude it is also very useful when locating sources of sound. Therefore the radiation patterns of complex vibrating machinery can be studied in situ.

Sound Fields

A sound field is a region where there is sound. It is classified according to the manner and the environment in which the sound waves travel. Some examples will now be described and the relationship between pressure and intensity discussed. This relationship is precisely known only in the first two special cases described below:

The Free Field

This term describes sound propagation in idealized free space where there are no reflections. These conditions hold in the open air (sufficiently far enough away from the ground) or in an anechoic room where all the sound striking the walls is absorbed. Free field propagation is characterized by a 6 dB drop in sound pressure level and intensity level (in the direction of sound propagation) each time the distance from the source is doubled. This is simply a statement of the inverse square law. The relationship between sound pressure and sound intensity (magnitude only) is also known. It gives one way of finding sound power which is described in the International Standard ISO 3745.

The Diffuse Field

In a diffuse field, sound is reflected so many times that it travels in all directions with equal magnitude and probability. This field is approximated in a reverberant room. Although the net intensity is zero, there is a theoretical relationship which relates the pressure in the room to the one-sided Intensity, Ix. This is the intensity in one direction, ignoring the equal and opposite component. One-sided intensity cannot be measured by a sound intensity analyzer but it is nevertheless a useful quantity: By measuring pressure we can use the relationship between pressure and one-sided intensity to find the sound power. This is described in ISO 3741.

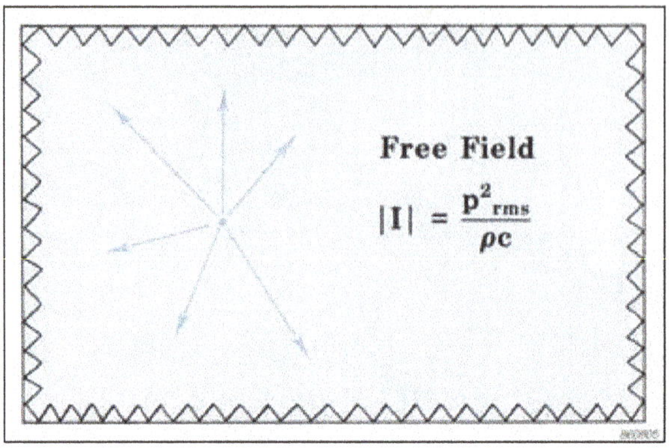

Free Field

$$|I| = \frac{p^2_{rms}}{\rho c}$$

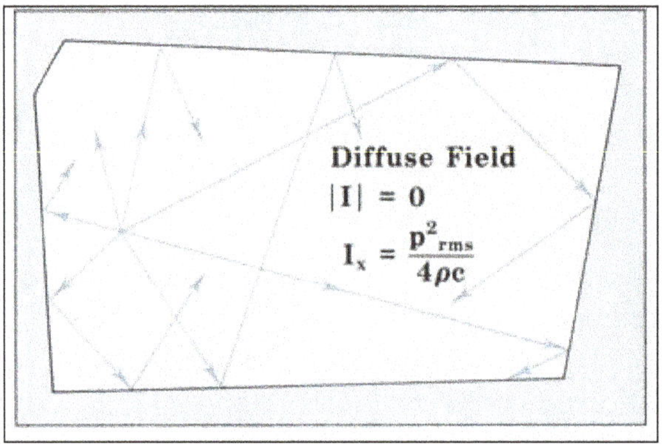

Diffuse Field

$$|I| = 0$$

$$I_x = \frac{p^2_{rms}}{4\rho c}$$

Active and Reactive Sound Fields

Sound propagation involves energy flow but there can still be a sound pressure even when there is no propagation. An active field is one where there is energy flow. In a pure reactive field, there is no energy flow. At any instant energy may be travelling outward, but it will always be returned at a later instant. The energy is stored as if in a spring. Hence the net intensity is zero. In general a sound field will have both active and reactive components. Pressure measurements for sound power in fields which are not well-defined can be unreliable, since the reactive part is unrelated to the power radiated. We can, however, measure sound intensity. Since sound intensity describes energy flow, there will be no contribution from the reactive component of the field. Two examples of reactive fields follow.

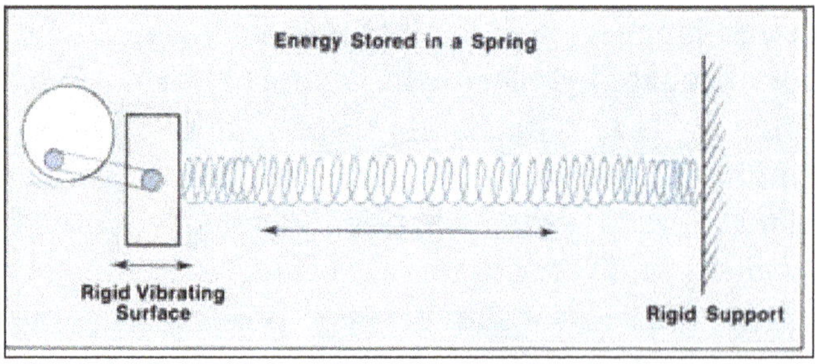

Energy Stored in a Spring

Rigid Vibrating Surface

Rigid Support

Standing Waves in a Pipe

Consider a piston exciting the air at one end of a tube. At the other end there is a termination which causes the sound waves to be reflected. The combination of the forward-travelling and reflected waves produces patterns of pressure maxima and minima which occur at fixed distances along the tube. If the termination is completely rigid all the energy is reflected and the net intensity is zero. With an absorptive termination some intensity will be measured. Standing waves are also present in rooms at low frequencies.

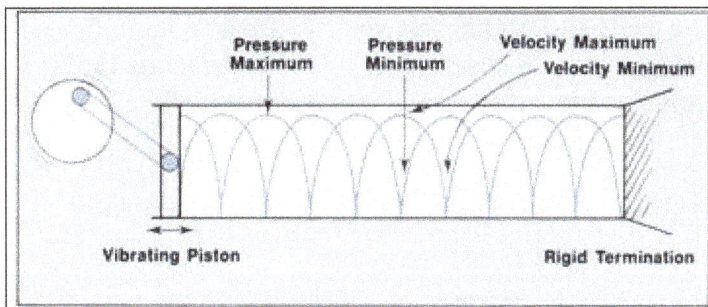

The Near Field of a Source

Very close to a source, the air acts as a mass-spring system which stores the energy. The energy circulates without propagating and the region in which it circulates is called the near field. Only sound intensity measurements for sound power determination can be made here. And because it is possible to get close to the source, the signalto-noise ratio is improved.

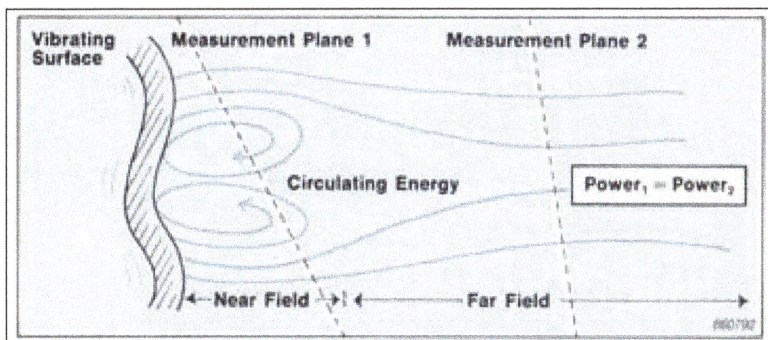

Pressure and Particle Velocity

When a particle of air is displaced from its mean position there is a temporary increase in pressure. The pressure increase acts in two ways: to restore the particle to its original position, and to pass on the disturbance to the next particle. The cycle of pressure increases (compressions) and decreases (rarefactions) propagates through the medium as a sound wave. There are two important parameters in this process: the pressure (the local increases and decreases with respect to the ambient) and the velocity of the particles of air which oscillate about a fixed position. Sound intensity is the product of particle velocity and pressure. And, as can be seen from the transformation below, it is equivalent to the power per unit area definition given earlier.

$$\text{Intensity} = \text{Pressure x Particle Velocity} = \frac{\text{Force}}{\text{Area}} \times \frac{\text{Distance}}{\text{Time}} = \frac{\text{Energy}}{\text{Area} \times \text{Time}} = \frac{\text{Power}}{\text{Area}}$$

In an active field, pressure and particle velocity vary simultaneously. A peak in the pressure signal occurs at the same time as a peak in the particle velocity signal. They are therefore said to be in phase and the product of the two signals gives a net intensity. In a reactive field the pressure and particle velocity are 90° out of phase. One is shifted a quarter of a wavelength with respect to the other. Multiplying the two signals together gives an instantaneous intensity signal varying sinusoidally about zero. Therefore the time-averaged intensity is zero.

In a diffuse field the pressure and particle velocity phase vary at random and so the net intensity is zero.

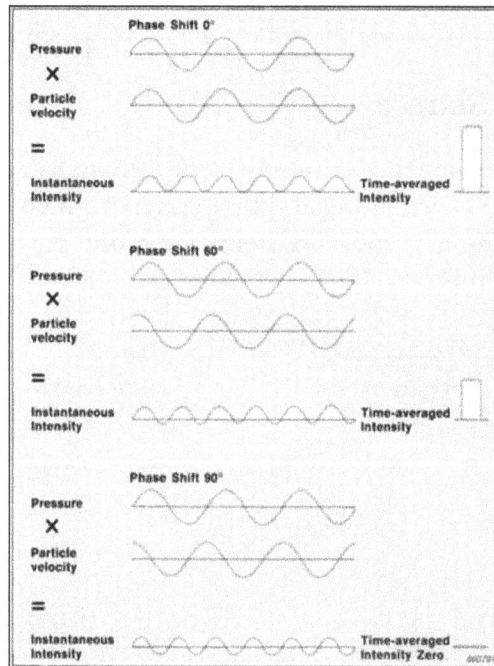

Measuring Sound Intensity

The Euler Equation: Finding the Particle Velocity

Sound intensity is the time-averaged product of the pressure and particle velocity. A single microphone can measure pressure — this is not a problem. But measuring particle velocity is not as simple. The particle velocity, however, can be related to the pressure gradient (the rate at which the instantaneous pressure changes with distance) with the linearized Euler equation. With this equation, it is possible to measure this pressure gradient with two closely spaced microphones and relate it to particle velocity.

Euler's equation is essentially Newton's second law applied to a fluid. Newton's Second Law relates the acceleration given to a mass to the force acting on it. If we know the force and the mass we can find the acceleration and then integrate it with respect to time to find the velocity.

$$F = ma$$

$$a = \frac{F}{m}$$

$$v = \int \frac{F}{m} dt$$

With Euler's equation it is the pressure gradient that accelerates a fluid of density p. With knowledge of the pressure gradient and the density of the fluid, the particle acceleration can be calculated. Integrating the acceleration signal then gives the particle velocity.

$$a = -\frac{1}{\rho} \text{grad p}$$

In one direction,

$$\frac{\partial u}{\partial t} = -\frac{1}{\rho} \frac{\partial p}{\partial r}$$

$$u = -\int \frac{1}{\rho} \frac{\partial p}{\partial r} dt$$

The Finite Difference Approximation

The pressure gradient is a continuous function, that is, a smoothly changing curve. With two closely spaced microphones it is possible to obtain a straight line approximation to the pressure gradient by taking the difference in pressure and dividing by the distance between them. This is called a finite difference approximation. It can be thought of as an attempt to draw the tangent of a circle by drawing a straight line between two points on the circumference.

The Intensity Calculation

The pressure gradient signal must now be integrated to give the particle velocity. The estimate of particle velocity is made at a position in the acoustic centre of the probe, between the two microphones. The pressure is also approximated at this point by taking the average pressure of the two microphones. The pressure and particle velocity signals are then multiplied together and time averaging gives the intensity.

A sound intensity analyzing system consists of a probe and an analyzer. The probe simply measures the pressure at the two microphones. The analyzer does the integration and calculations necessary to find the sound intensity. These equations are not new. What is new is the use of modern signal processing techniques to implement the equation. This can be done in two ways: by directly using integrators and filters (analogue or digital) to implement the equation step by step, or by using an FFT analyzer. The latter relates the intensity to the imaginary part of the cross spectrum (a mathematical term) of two microphone signals. The formulations are equivalent; both give the sound intensity.

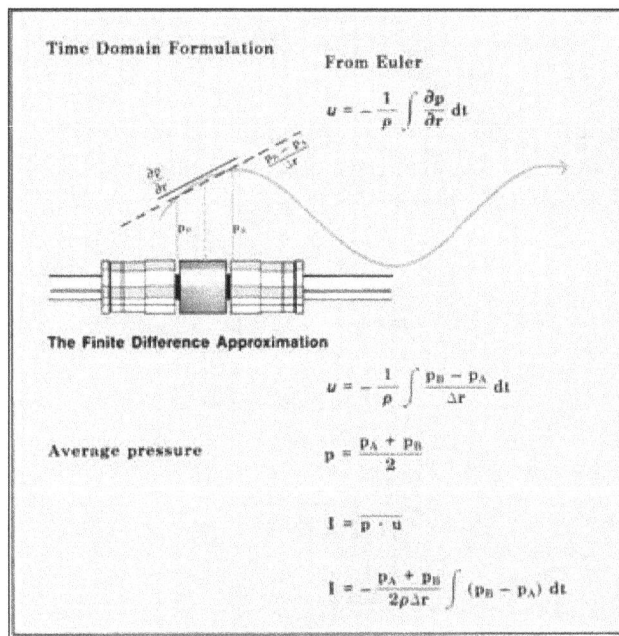

Frequency Domain Formulation for FFT Analyzers,

$$I = -\frac{1}{\rho\omega\Delta r}\,\mathrm{Im}\,G_{AB}$$

ω is the angular frequency,

$\mathrm{Im}\,G_{AM}$ is the imaginary part of the cross spectrum.

The Sound Intensity Probe

The Brüel and Kjær probe has two microphones mounted face to face with a solid spacer in between. This arrangement has been found to have better frequency response and directivity characteristics than side-by-side, back-to-back or face-to-face without solid spacer arrangements. Three solid spacers define the effective microphone separation to 6, 12 or 50 mm. The choice of spacer depends on the frequency range to be covered.

Half-inch microphones are used for lower frequencies. But smaller quarter-inch micro-phones are used at high frequencies to reduce interference effects.

Directivity Characteristics

The directivity characteristic for the sound intensity analyzing system looks (two-dimensionally) like a figure-of-eight pattern — known as a cosine characteristic. This is due to the probe and the calculation within the analyzer.

Since pressure is a scalar quantity, a pressure transducer should have an equal response; no matter what the direction of sound incidence (that is, we need an omnidirectional characteristic). In contrast, sound intensity is a vector quantity. With a two-microphone probe, we do not measure the vector however; we measure the component in one direction, along the probe axis. The full vector is made up of three mutually perpendicular components (at 90° to each other) — one for each coordinate direction.

For sound incident at 90° to the axis there is no component along the probe's axis, as there will be no difference in the pressure signals. Hence there will be zero particle velocity and zero intensity. For sound incident at an arbitrary angle 6 to the axis the intensity component along the axis will be reduced by the factor $\cos \theta$. This reduction produces the cosine directivity characteristic.

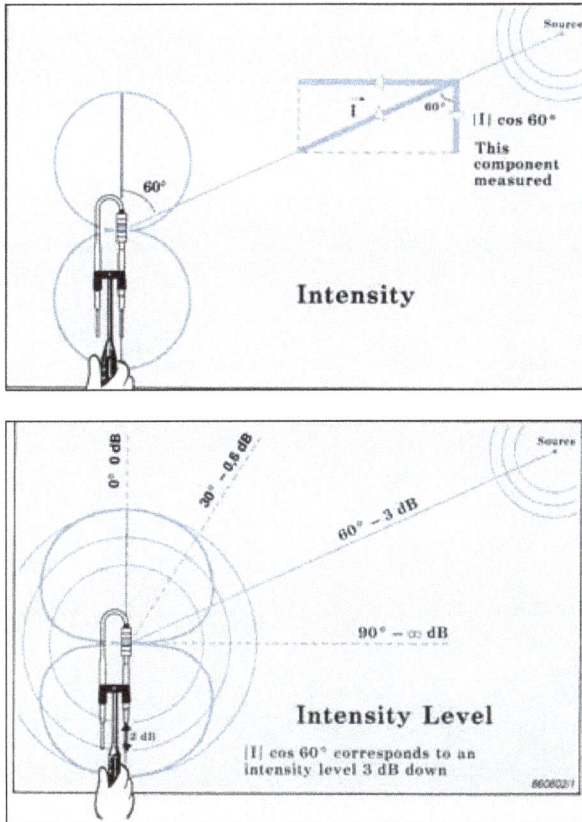

Reference Levels

The sound pressure, intensity, power and particle velocity levels, (L_p, L_I, L_w and L_u respectively), are all measured in dBs. Decibels are a ratio of the specified quantity measured against some reference. For pressure the reference level is chosen so that it corresponds approximately to the threshold of hearing.

Other reference levels have been approximately related to this by using the free field relations between pressure and intensity, and pressure and particle velocity. And in the free field we will obtain the same dB reading irrespective of whether we measure pressure, intensity or particle velocity (measured in the direction of propagation). Actually, because round numbers have been chosen for the reference levels, there is a slight difference in levels. The actual difference depends on the value of the characteristic impedance, ρc, of the medium in which it is measured. Here ρ is the density and c the speed of sound in the medium. The difference is usually negligible in air except at high altitudes. To avoid possible confusion with pressure levels, sound power levels are sometimes given in bels — 10 dB equals 1 bel.

In the free field the pressure and intensity levels in the direction of propagation are numerically the same. However, intensity measurements in the free field are not needed. In practice, we will not measure in a free field and so there will be a difference between the pressure and intensity levels. This difference is an important quantity known as the pressure-intensity index (previously known as the phase index or reactivity index with different sign).

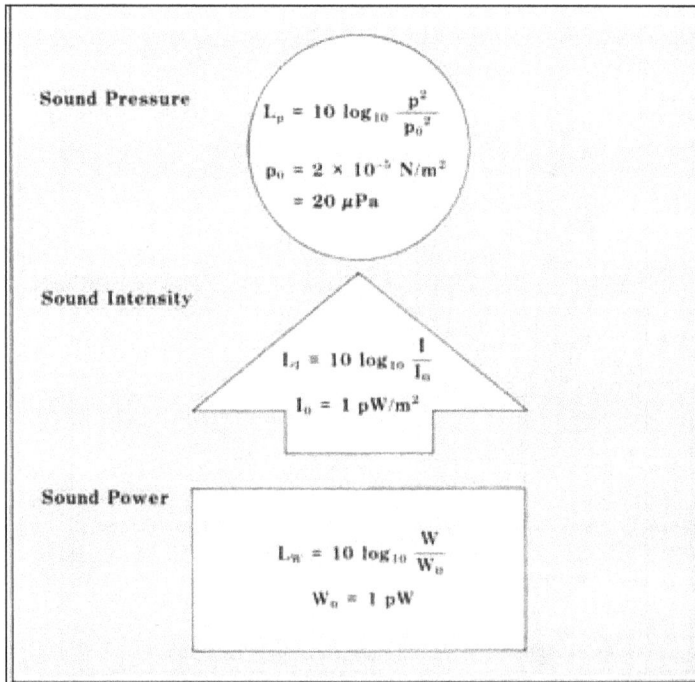

Sound Pressure

$$L_p = 10 \log_{10} \frac{p^2}{p_0^2}$$

$$p_0 = 2 \times 10^{-5} \text{ N/m}^2$$

$$= 20 \ \mu\text{Pa}$$

Sound Intensity

$$L_I = 10 \log_{10} \frac{I}{I_0}$$

$$I_0 = 1 \ \text{pW/m}^2$$

Sound Power

$$L_w = 10 \log_{10} \frac{W}{W_0}$$

$$W_0 = 1 \ \text{pW}$$

For Free Field Radiation $I = \dfrac{p^2_{rms}}{\rho c}$

If $\rho c = 400 \ Nsm^{-3} \quad L_p = L_1 = L_u$

$\rho c = 415 \ Nsm^{-3}$ at $20°$ C and 1013 hPa

$\therefore L_I = L_p - 0.16$ dB

Permissions

All chapters in this book are published with permission under the Creative Commons Attribution Share Alike License or equivalent. Every chapter published in this book has been scrutinized by our experts. Their significance has been extensively debated. The topics covered herein carry significant information for a comprehensive understanding. They may even be implemented as practical applications or may be referred to as a beginning point for further studies.

We would like to thank the editorial team for lending their expertise to make the book truly unique. They have played a crucial role in the development of this book. Without their invaluable contributions this book wouldn't have been possible. They have made vital efforts to compile up to date information on the varied aspects of this subject to make this book a valuable addition to the collection of many professionals and students.

This book was conceptualized with the vision of imparting up-to-date and integrated information in this field. To ensure the same, a matchless editorial board was set up. Every individual on the board went through rigorous rounds of assessment to prove their worth. After which they invested a large part of their time researching and compiling the most relevant data for our readers.

The editorial board has been involved in producing this book since its inception. They have spent rigorous hours researching and exploring the diverse topics which have resulted in the successful publishing of this book. They have passed on their knowledge of decades through this book. To expedite this challenging task, the publisher supported the team at every step. A small team of assistant editors was also appointed to further simplify the editing procedure and attain best results for the readers.

Apart from the editorial board, the designing team has also invested a significant amount of their time in understanding the subject and creating the most relevant covers. They scrutinized every image to scout for the most suitable representation of the subject and create an appropriate cover for the book.

The publishing team has been an ardent support to the editorial, designing and production team. Their endless efforts to recruit the best for this project, has resulted in the accomplishment of this book. They are a veteran in the field of academics and their pool of knowledge is as vast as their experience in printing. Their expertise and guidance has proved useful at every step. Their uncompromising quality standards have made this book an exceptional effort. Their encouragement from time to time has been an inspiration for everyone.

The publisher and the editorial board hope that this book will prove to be a valuable piece of knowledge for students, practitioners and scholars across the globe.

Index